Uncertainties in Acoustical Transfer Functions

Modeling, Measurement and Derivation of Parameters for Airborne and Structure-borne Sound

Von der Fakultät für Elektrotechnik und Informationstechnik der
Rheinischen-Westfälischen Technischen Hochschule Aachen
zur Erlangung des akademischen Grades eines

DOKTORS DER INGENIEURWISSENSCHAFTEN

genehmigte Dissertation

vorgelegt von

Diplom-Ingenieur

Pascal Dietrich

aus Dorsten, Deutschland

Berichter:

Universitätsprofessor Dr. rer. nat. Michael Vorländer
Universitätsprofessor Dr.-Ing. Kay Hameyer

Tag der mündlichen Prüfung: 11. Oktober 2013

Diese Dissertation ist auf den Internetseiten der Hochschulbibliothek online verfügbar.

Pascal Dietrich

Uncertainties in Acoustical Transfer Functions

**Modeling, Measurement and Derivation of Parameters
for Airborne and Structure-borne Sound**

Logos Verlag Berlin GmbH

λογος

Aachener Beiträge zur Technischen Akustik

Editor:
Prof. Dr. rer. nat. Michael Vorländer
Institute of Technical Acoustics
RWTH Aachen University
52056 Aachen
www.akustik.rwth-aachen.de

Bibliographic information published by the Deutsche Nationalbibliothek

The Deutsche Nationalbibliothek lists this publication in the Deutsche Nationalbibliografie; detailed bibliographic data are available in the Internet at http://dnb.d-nb.de .

D 82 (Diss. RWTH Aachen University, 2013)

ISBN 978-3-8325-3551-3
ISSN 1866-3052
Vol. 16

Logos Verlag Berlin GmbH
Comeniushof, Gubener Str. 47,
D-10243 Berlin
Tel.:   +49 (0)30 / 42 85 10 90
Fax: +49 (0)30 / 42 85 10 92
http://www.logos-verlag.de

Contents

Glossary

Acronyms

ADC	Analog to Digital Converter
AR	Auto Regression
ARMA	Auto Regression Moving Average
AWGN	Additive White Gaussian Noise
BEM	Boundary Element Simulation
CAPZ	Common Acoustic Poles and Zeros approach
CMRR	Common Mode Rejection Ratio
DAC	Digital to Analog Converter
DFT	Discrete Fourier Transform
DUT	Device Under Test
EDT	Early Decay Time
FEM	Finite Element Simulation
FFT	Fast Fourier Transform
FHT	Fast Hadamard Transform
FIR	Finite Impulse Response
FRF	Frequency Response Function
GUM	Guide to the expression of Uncertainty in Measurement
IIR	Infinite Impulse Response
IR	Impulse Response
ISM	Image Source Method
ITA	Institute of Technical Acoustics at RWTH Aachen University
JND	Just Noticeable Difference
LSB	Least Significant Bit
LTI	Linear Time-Invariant

MA	Moving Average
MC	Monte Carlo
MESM	Multiple Exponential Sweep Method
MIMO	Multiple Input Multiple Output
MLS	Maximum Length Sequence
OOP	Object-Oriented Programming
OTPA	Operational Transfer Path Analysis
PDF	Propability Density Function
QUIESST	QUIetening the Environment for a Sustainable Surface Transport
RI	Reflection Index
RIR	Room Impulse Response
RT	Reverberation Time
SH	Spherical Harmonics
SNR	Signal-to-Noise Ratio
TF	Transfer Function
THD	Total Harmonic Distortion
TP	Transfer Path
TPA	Transfer Path Analysis
TPS	Transfer Path Synthesis

Notation

x	scalar, real-valued or complex-valued
$\mathbf{x}$	vector, real-valued or complex-valued
$x(t)$	function over time t (time domain)
$X(f)$, $X(\omega)$	function over frequency f (frequency domain)
$s(t) * h(t)$	convolution in time domain
$S(f) \cdot H(f)$	multiplication in frequency domain
$\mathcal{F}$	Fourier transformation
$(\ldots)_{\mathrm{r}}$	subscript for receiver
$(\ldots)_{\mathrm{s}}$	subscript for source
$(x, y, z, \ldots)$	vector with elements x, y, z, $\ldots$
$(\ldots)_{\mathrm{T}}$	subscript for translational components or submatrix
$(\ldots)_{\mathrm{R}}$	subscript for rotational components or submatrix
$\mathrm{diag}\,(\ldots)$	diagonal of matrix

$(\ldots)^T$	transpose of matrix
$\mathbf{I}$	identity matrix

List of Symbols

$j = \sqrt{-1}$	the imaginary unit
e	the EULER constant, exponential function
i, j	indices, integer values
t	time in s
f	frequency in Hz
$\omega = 2\pi f$	angular frequency in $\mathrm{rad/s}$
c	speed of sound (in air at $20°$C: $c = 343\,\mathrm{m/s}$)
$\lambda = c/f$	wave length in m
$k = 2\pi/\lambda = \omega/c$	wavenumber in $1/\mathrm{m}$
$s(t), S(f)$	abstract input signal
$g(t), G(f)$	abstract output signal
$h(t)$	impulse response (time domain)
$H(f)$	frequency response (frequency domain)
$n(t), N(f)$	noise signal or noise spectrum
p	sound pressure in Pa
v	velocity or particle velocity in $\mathrm{m/s}$
F	mechanical force in N
M	mechanical moment in Nm
o	angular velocity in $°/\mathrm{s}$
a	acceleration in $\mathrm{m/s^2}$
ξ	displacement in m
Y	mobility or admittance
Z	impedance
R	reflection factor
ϕ, θ	angle or phase in $°$ or rad
ψ	eigenfunction for room modes
L	room dimensions in m
V	volume in $\mathrm{m^3}$
f_M	uncertainty model function
k	harmonic or polynomial order in context of nonlinearities
τ	time difference or delay
$\ln, \log, \log_2$	logarithms to the base e, 10 and 2

Abstract

Measured transfer functions of acoustic systems are often used to derive single-number parameters. The uncertainty analysis is commonly focused on the derived parameters but not on the transfer function as the primary quantity. This thesis presents an approach to assess the independent uncertainty contributions in these transfer functions by using analytic models to finally provide a detailed uncertainty analysis. Firstly, uncertainties caused by the measurement method are analyzed with a focus on the underlying signal processing. In particular, the influence of nonlinearities in the acoustic measurement chain are modeled to predict artifacts in the measured signals and hence the calculated acoustic transfer function. Secondly, characterization methods commonly applied in field of signal processing are linked to the acoustic scenarios and the main influencing parameters. Acoustic parameters are then derived analytically and by means of Monte Carlo simulations considering the uncertainty of these input parameters.

In order to provide airborne applications, analytic models for sound barrier and room acoustic measurements are developed incorporating the directivity and the orientation of the sound source as well as the position of sources and receivers. Furthermore, a measurement method to predict the influence of arbitrary directivities on the transfer function in a post-processing step is developed. The simulated uncertainty contributions are successfully validated by measurements. Additionally, the coupling between the sound source and the receiver in structure-borne scenarios is investigated theoretically since this coupling has generally a stronger influence on the transfer function then for airborne scenarios. Results obtained by three different measurement methods are compared and their principal differences are investigated. Moreover, an application incorporating an analytic plate model is exemplarily used to predict the uncertainties due to inaccuracies in the sensor positions similar to the airborne applications. Finally, the influence caused by simplifications applied to the exact model in terms of neglecting cross-coupling is analyzed for this scenario.

1

Introduction

The assessment of the range of uncertainty of a given measurement quantity is as important as the determination of the measured value itself. Mainly single number values are considered in this context. The error calculus has been summarized along with a guideline how to obtain the measurement uncertainty in a unified manner in the Guide to the expression of uncertainty (GUM) (JCGM 100, 1995). Many recent measurement standards refer to this guideline and claim for the assessment of the uncertainty of the defined quantity, e.g., the standard defining the measurement of room acoustic parameters ISO 3382 (ISO 3382, 2009). Especially for measurement quantities that are calculated based on single number parameters with known measurement uncertainties the assessment by means of the guideline is fairly simple. Problems arise if the primary measurement quantities are not available as single number parameters but as a set of several hundred values with certain correlation and hence containing redundancy instead.

As a consequence so called round-robin tests for measurements, simulation programs or evaluation procedures are conducted world wide to obtain the standard deviations between the different groups and predict the uncertainty. Although this is a practical approach complying with the GUM rules it has the drawback in being expensive in time and hence costs and delivering only results based on statistical deviations including all factors influencing the deviation between the groups of the measurement at once without yielding information on systematic errors or the main causes of these uncertainties.

In the field of acoustics the sound pressure level is a well-known example of a single number parameter that is obtained by analysis of time signals. But there are recent standards of acoustic parameters, e.g., defining the reflection index of sound barriers or the room acoustic parameters, that are all based on measured transfer functions of acoustic systems or specimen. These acoustic

1

transfer functions are commonly represented by discrete sampled values in either frequency or time domain that cannot be handled as a single value parameter.

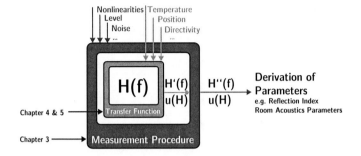

Figure 1.1.: Modeling approach used in this thesis to investigate the measurement uncertainties divided into uncertainties introduced by the measurement of transfer functions and uncertainties caused by perturbations of boundary conditions of the system to be measured.

Application that incorporate the determination of acoustic transfer functions range from, e.g., laboratory measurements of loudspeakers, microphones, acoustic barriers or absorbers over room acoustic measurements in the field up to the characterization of vibration paths in structures and also the acoustic radiation of these structures. Although the purpose of the measurement results and the representation of these results strongly depend on the application, the basic concepts of the determination of these transfer functions are very similar. In room acoustics, the transfer function is merely used to derive single number values in the time domain to rate and compare the acoustic quality of, e.g., concert halls or auditoria. For troubleshooting in, e.g., the automotive industry transfer functions or frequency responses are analyzed in terms of resonance frequencies, damping and sound transmission. For applications in condition monitoring, e.g., changes in these transfer functions are analyzed to detect a drift or ageing of structures. In the field of auralization, transfer functions are directly used to be filtered with dry source signals to listen to acoustic scenarios yielding a powerful tool for the assessment of the sound quality of products, e.g, involving listening tests. Although the transfer function is often considered as a measurement result it could be also substituted by simulation results obtained by different simulation methods. In combination with the auralization approach this leads to the acoustical virtual prototyping but it enables a new approach to the uncertainty analysis as well.

In some of these applications the measurement uncertainties in the derived parameters were already addressed but these publications do not investigate the uncertainties in the primary measured quantity, which is the transfer function. A different approach combines measurement data with Monte Carlo simulations as, e.g. published in (E. Brandão et al., 2011; E. Brandão, Lenzi, and Cordiolli, 2011) to determine the measurement uncertainty of the in-situ absorption coefficient. As the measurement method comprises of a measurement of a transfer function and a post-processing step involving the geometry of the measurement setup, the influence of the uncertainties in the geometric data are considered by means of Monte Carlo simulations yielding the uncertainty contribution in the derived parameter. As a drawback, this practical approach does not incorporate all uncertainties and is especially not capable of detecting systematic errors. For the same applications the uncertainty was also addressed by means of extensive numerical simulations of the measurement setup (Dietrich et al., 2012; Hirosawa et al., 2009; Müller-Trapet et al., 2013; Otsuru et al., 2009) yielding additionally insight on the systematic errors and the separate uncertainty contributions. In the field of room acoustics, different simulation approaches to assess the uncertainties in room acoustic parameters are currently used by different researchers. These methods range from modifications of measured room impulse responses, over numerical finite element or boundary element simulation results to responses obtained by ray tracing methods. Each method is considered to deliver reliable results and the applicability for the uncertainty analysis especially by means of Monte Carlo simulations and the choice of the method is merely a matter of computation time and frequency range. Ready-to-use concepts on how to model uncertainties in complex transfer functions and especially acoustic transfer functions have not been found in the literature although there is an increasing number of publications dealing with uncertainties especially in structure-borne sound modeling taking the transfer function increasingly into consideration (Evans, 2010).

The present thesis aims at giving a comprehensive insight into the uncertainties in measurements of acoustic transfer functions by means of parametric models as depicted in Figure 1.1 in an abstract manner. Main influencing quantities are separated in the ones having an influence on the underlying transfer function $H(f)$ in terms of its uncertainty $u(H)$, e.g. temperature, position or directivty and the ones having an influence on the quality of the measurement procedure, e.g., nonlinearities and noise. The schematic blocks are characterized by means of analytic models. Moreover, these models are then used for either analytic evaluation or Monte Carlo simulations along with given uncertainties of the input parameters in order to investigate potential errors or the uncertainty of these transfer functions. Single number parameters are then derived. The central idea is to develop detailed parametric models of the measurement or the transfer function where the parameters are then treated as single number parameters with uncertainties. Hence, the calculus of the GUM becomes applicable and a simple to use simulation approach for the assessment of uncertainties in transfer functions is obtained.

Outline of the present thesis This thesis is structured as follows with the aim of separating the common basics, the measurement procedure itself and the transfer functions. Chapter 2 summarizes the basics from signal theory, acoustics and uncertainty modeling including important considerations for the further processing of data. Following are three main chapters exploring sources of uncertainties in three different fields. Each chapter introduces the theoretic background required for the particular modeling approaches separately. In Chapter 3 the measurement procedure of acoustical transfer function is investigated with a special focus on the artifacts caused by signal processing. The effects of nonlinear systems in the measurement chain is studied in an abstract manner whereas an application example for the uncertainties in room acoustic measurements is presented. The transfer functions for airborne and structure-borne sound are investigated separately. Chapter 4 focuses on the parametric description of airborne transfer functions by analytic models and based on measurements. The influence of positioning uncertainties of sources and receivers is investigated in a general perspective and in particular for three application examples. In Chapter 5 the coupling of structure-borne sound sources and receivers is studied. Based on these findings the main transfer path analysis methods are characterized. Finally, an application example using an analytic model of a structure-borne system is used to study the uncertainties in a similar manner as for the airborne sound scenarios. Details on implementation, mathematical deductions and additional uncertainty simulations are presented in the Appendix.

2

Fundamentals

This chapter gives a brief overview of the quantities and conventions used in the field of acoustics. An overview of the fundamentals in signal processing required for acoustic measurements is presented and the concept of transfer functions is introduced along with modeling techniques. The combination of the basics with their special focus on acoustics delivers new insight in terms of causes of uncertainties. Finally, the modeling and calculation of the uncertainty of physical quantities is briefly introduced.

2.1. Physical and Acoustical Quantities – Conventions

Acoustics and vibration are interdisciplinary topics as, e.g., noise problems force engineers to develop silent and euphonic products as customers also rate these products such as house-hold appliances or cars by their acoustic properties. For multimedia or pro-audio equipment the entire sound propagation should deliver high sound pressure levels for a broad frequency range with low distortion at the same time. Since experts with different backgrounds use different terminologies it is essential to specify the conventions and the naming used in this work. The main groups working on this topic are experts from electrical engineering, signal processing, mechanical engineering and physics. This thesis focuses on the audible frequency range of 20 Hz to 20 kHz that is used quite commonly in acoustics. However, results might be directly applicable to other frequency ranges as well.

According to KUTTRUFF "Acoustics is the science of sound and deals with the origin of sound and its propagation, either in free space, or in pipes and channels, or in closed spaces." (KUTTRUFF, 2007). He implicitly refers to airborne sound propagation. Vibration can be associated with structural dynamics and mechanical oscillations. FAHY AND GARDONIO use the term *vibroacoustic* instead (F. FAHY and GARDONIO, 2007). In this thesis, the terms *airborne* and *structure-borne* are used to distinguish between these two different modes of sound propagation. The term acoustical transfer path is used for both airborne and structure-borne transfer paths and also combination of both. The following short definitions are based on (KUTTRUFF, 2007), (MECHEL, 2008) and (F. FAHY and GARDONIO, 2007).

With the term *airborne sound* the propagation of information by waves in a fluid—especially in air—is summarized. Only longitudinal waves are observed in fluids. The acoustic field quantities used are the *sound pressure p* and the *particle velocity v*. In general, the sound pressure is a scalar and the particle velocity a three-dimensional vector and therefore written in bold letters as $\mathbf{v}$. These quantities are also called primary acoustic field quantities. The sound pressure is of major interest for noise control since human ears are pressure receivers. This is explained as it is simple to measure the sound pressure directly for a wide frequency range from a practitioner's point of view. The particle velocity can be calculated from the pressure gradient by using two or more pressure microphones or by using anemometers such as the *Microflown Technologies pu-probe*. Complex-valued quantities are not marked as such for improved readability in this thesis.

These primary quantities can be directly observed as time varying signals which are specifically written as, e.g., $p(t)$ at some points. The particle displacement ξ can be calculated from the particle velocity by integration over time and the particle acceleration $\mathbf{a}$ by time differentiation. The ratio of sound pressure and particle velocity is called the field impedance $Z = p/v$ where only one specific component of the velocity vector is used. The sound intensity $\mathbf{I}$ is defined as the product of the sound pressure and the conjugate complex particle velocity vector and is, e.g., used to calculate the sound power P.

The term *structure-borne sound* covers the propagation of sound or vibration energy in solid structures. Both longitudinal and transversal waves occur (F. FAHY and GARDONIO, 2007). For better readability, no distinction between airborne and structure-borne quantities is considered in the notation in terms of subscripts. Generally, the physical quantities are the force F and the velocity v, similar as for the airborne sound. But in contrast, the force and the velocity are also vectors $\mathbf{F}$ and $\mathbf{v}$, respectively. In case not only translational but also rotational components become prominent these vectors consist of six dimensions, denoted as *degrees of freedom* in the following. These vectors are used in a general notation combining translational components (force or velocity) in x,

y and z-direction and rotational components (moments or angular velocity) around the x, y and z-direction in the following. The force vector reads as $\mathbf{F} = (F_x, F_y, F_z, M_x, M_y, M_z)^{\mathrm{T}}$ and the velocity vector as $\mathbf{v} = (v_x, v_y, v_z, o_x, o_y, o_z)^{\mathrm{T}}$. In case more than one contact point is described the force vectors of the single contact points $\mathbf{F_i}$ can be summarized in a vector $\mathbf{F} = (\mathbf{F}_1^{\mathrm{T}}, \mathbf{F}_2^{\mathrm{T}}, \ldots)^{\mathrm{T}}$. With the definition used the number of degrees of freedom does not scale with the number of contact points considered.

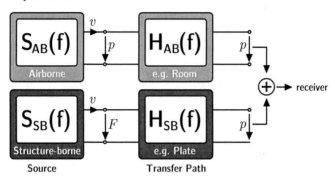

Figure 2.1.: Schematic block diagram for transfer path analysis and synthesis for structure-borne and airborne sound propagation including the coupling between source and receiver [based on (Vorländer, 2007)].

From a signal processing point of view the analysis of an airborne or structure-borne acoustic system is based on the same model as, e.g., for the ongoing synthesis of signals with for the same scenario. This is illustrated in Figure 2.1 in a simple manner for only one exemplary airborne and one structure-borne path. However, practical issues in measurement and the goal of experimenter forces to distinguish. Transfer Path Analysis (TPA) is the term used for several methods with the aim to determine the individual Transfer Paths (TPs) from sound sources to one or more receivers. The measurement of the source signals can be associated with this task. Transfer Path Synthesis (TPS) aims at superposing source signals with the corresponding TPs to obtain the signal at the receiving position. Since both the amplitude of the source signal and the amplitude of the TP are important for the amplitude observed at the receiver the *path contribution* as the convolution of the source signal and TP can give more detailed information. The latter is part of a TPS.

The term *auralization* is used when path contributions, virtual scenarios or other data is converted to time signals and playback, e.g., by loudspeakers. A detailed definition along with several methods is given in (Vorländer, 2007).

2.2. Basics in Signal Processing

(OPPENHEIM, 1996) provides detailed information on mainly Linear Time-Invariant (LTI) system theory, time-continuous and time-discrete signals and systems. A detailed overview of system theory, system models and signal processing background with a special focus on acoustic systems is presented in (TOHYAMA and KOIKE, 1998). Both references are exploited in this chapter and linked to the acoustic problem.

The relation between the time and the frequency domain is essential in this thesis since the domain of signals is frequently switched for calculation and illustration purposes. All signals are primarily measured in the time domain and represented digitally within any state-of-the-art measurement device. Hence, effects of quantization and the sampling theorem have to be considered. Filtering in the frequency domain, e.g., band-pass, high-pass, low-pass, have impact in the time domain observed, e.g., as broadening of the Impulse Response (IR) and on the other hand, windowing in the time domain has an impact on the frequency response, e.g., observed as smoothing as well. The windowing technique is used to extract information from measured impulse responses by separating the early part containing the impulse response information from the late part containing noise, unwanted reflections or other artifacts (MÜLLER and MASSARANI, 2001).

2.2.1. Description of Systems and the Time-Frequency Relation

A *signal* is a variable represented in either the time or the frequency domain. It can contain noise, speech, music, impulses or arbitrary information. It is of interest to analyze the frequency spectrum of a signal since our ears analyze sounds according to their frequency content as well. The Fourier transform provides a link between the signal $s(t)$ in *time domain* and the spectrum $S(f)$ in *frequency domain* by the Fourier integral for continuous signals (OPPENHEIM, 1996). Since the integration time is infinite this transform is only applicable in theoretic considerations. In the field of acoustics the signals $s(t)$ are only real-valued leading to a symmetry in frequency domain

$$S(f) = S^*(-f). \tag{2.1}$$

Due to this symmetry the information on the negative frequency axis is redundant and can be neglected. Hence, plots of spectra in this thesis will only show the positive frequency axis.

The term *system* is used as a summarizing container for arbitrary transfer elements and is similar to the *black-box* approach as, e.g., used in (VORLÄNDER, 2007). In this thesis, systems are assumed to always consist of at least one input and at least one output. LTI systems can be completely described by their causal, real-valued impulse response $h(t)$ or their complex-valued transfer function $H(f)$. The signal at the output $g(t)$ is the convolution of the input signal and the impulse response $g(t) = h(t) * s(t)$.

The linearity requirement enables the prediction of the output signal due to an amplitude change at the input by a factor as with a change in the output by the same factor. Furthermore, the superposition approach is applicable. Time-invariance refers to a system that does not change its behavior over time. Both assumptions have to be critically considered when it comes to practical measurements.

2.2.2. Sampling in Time Domain

Signal processing, acquisition and analysis can be efficiently realized by means of digital signal processors and processing. Therefore, the influence of sampling or discretization in the time domain has to be considered. Only equidistant time sampling is considered in this thesis. The sampling rate or sampling frequency f_s defines the distance between two samples in the time domain as $\Delta T = 1/f_s$. This can be modeled by a multiplication of the signal $s(t)$ with an impulse train in the time domain and results in an infinite repetition of the spectrum $S(f)$ with the distance $\Delta f = f_s$. In case $S(f)$ has information for frequencies above the Nyquist or Shannon frequency $f_{NY} = f_s/2$ it also has information below $-f_{NY}$ due to the symmetry. Hence, this repetition results in a superposition of the original spectrum with the repeated spectrum or alias. The effect is called (frequency domain) *aliasing*. This sampling theorem has to be fulfilled each time an analog signal is sampled by an Analog to Digital Converter (ADC) or a continuous signal representation, e.g., inside a computer program, is discretized. The latter is less obvious and might lead to implementation errors. Due to aliasing and the repetition only information up to the Nyquist frequency has to be stored since the rest is redundant. Figure 2.2 illustrates the discretization effects.

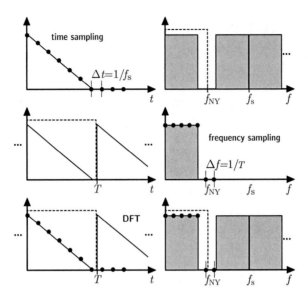

Figure 2.2.: The effects of discretization in both time and frequency domain. Top: Time discretization, Middle: Frequency domain discretization, Bottom: Discretization in both domains, equivalent to DFT.

2.2.3. Sampling in Frequency Domain

Sampling or discretization in the frequency domain can be analyzed in the same manner as in the time domain, except the continuous frequency domain is sampled with a fixed distance of $\Delta f = 1/T$. This directly results in a repetition of the original time signal with a shift of T. In case the original time signal is not zero outside this block with the size T aliasing in the time domain occurs. This T is analog to the sampling frequency. Only information inside of this block is usually to stored to avoid redundancy.

The combination of both of these sampling artifacts is equivalent to the basic behavior observed in the DFT, where a finite number of equidistant *time samples* are transformed to finite number of *frequency bins*. In computer programs an efficient algorithm, the Fast Fourier Transform (FFT), is used to speed up calculations. Hence, all simulated or measured signals analyzed in this thesis might show an influence of either time or frequency domain sampling.

2.2.4. Discrepancy between Time and Frequency Domain Models

Since digital data acquisition systems use sampling in the time domain, problems might occur when combining this data with frequency sampled data. In case an analytic expression of the impulse response exists this could be directly transformed to the frequency domain and it could be discretized in either domain. However, the sampling of this expression in the time domain and ongoing discrete transformation to the frequency domain can yield different results than frequency domain sampling of the continuously transformed result. Figure 2.3 shows these differences by using a single resonator with a specific frequency that can be expressed analytically in terms of its impulse response or frequency response. The phase of the frequency response always tends to either 0 or 180° towards the Nyquist frequency when time sampling is used. However, the phase of the frequency sampled data follows the exact formulation. Deviations in the magnitude are observed at both ends of the frequency axis. In general, time domain sampling of a causal continuous formulation yields a causal impulse response, whereas the frequency sampling could result in a rising in the end of the impulse response to maintain the correct spectrum. Both solutions are only approximations and the artifacts have to be always considered. An increase of T for the time sampling and of f_s for the frequency sampling can decrease the artifacts.

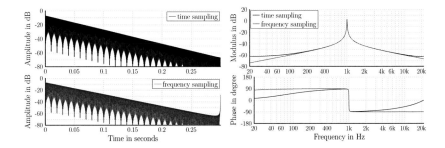

Figure 2.3.: Comparison of impulse response (left) and frequency response (right) with amplitude (top) and phase (bottom) of a single resonator ($f = 1\,\text{kHz}$, reverberation time RT = 0.3 s) with sampling applied in the time domain and frequency domain.

In practical examples, both forms of sampling or even a combination of both can occur. Frequency domain simulations, e.g., Finite Element Simulation (FEM) or Boundary Element Simulation (BEM) usually evaluate physical scenarios at discrete frequency points only. Additionally, the number of frequency points is finite resulting in a frequency band limitation. The transformation of these results to time domain directly results in acausal impulse responses although the simulations are correct for all frequencies as a real world scenarios cannot show acausal behavior. For auralization purposes this behavior is troublesome as the rising impulse in the end can become audible as an unwanted echo and hence should be avoided.

A practical example is the time integration or differentiation of signals, e.g., to obtain the acceleration from the measured velocity. For harmonic signals, both operations result in a trivial multiplication by $j\omega$ or its inverse respectively. The continuous function $j\omega$ is usually sampled in the frequency domain and acausality occurs. E.g., in the field of structure-borne sound modeling analytic expression of impedances usually sampled in the frequency domain.

The error in the time domain caused by frequency sampling ($\Delta f = 1/T$) compared to the time sampled version can be approximated by assuming an exponential decay of the energy with $e^{-t/\tau}$ and $\tau = RT/3\ln(10)$. RT stands for the reverberation or decay time of the resonator. The energy of the time-aliasing components in relation to energy of the time sampled signal with finite length T is calculated as

$$\varepsilon_{ft}(T) = \frac{\int\limits_0^T e^{-t/\tau}\,dt}{\int\limits_T^\infty e^{-t/\tau}\,dt} = 1 - e^{T/\tau}. \tag{2.2}$$

This relation is calculated logarithmically in dB in Figure 2.4 and might me considered as a *signal-to-aliasing-ratio* or Signal-to-Noise Ratio (SNR) due to time aliasing. Typical SNRs for auralization purposes are about $60 - 80\,\mathrm{dB}$ (VORLÄNDER, 2007) and hence the frequency resolution should be at least $\Delta f = 1.3/RT$ or $\Delta f = 1/RT$, respectively.

As a conclusion, the correct formulation in the continuous domain might result in serious problems after sampling. It seems reasonable to apply additional sampling if necessary only in the same domain as of the final result. Otherwise, causality has to be ensured by post-processing, e.g., time windowing.

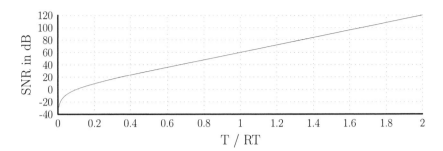

Figure 2.4.: Ratio between signal energy and energy of time-aliasing components for exponential energy decay with reverberation time RT by frequency sampling with $\Delta f = 1/T$.

2.2.5. Quantization in Time Domain

Quantization or the discretization of the values is necessary as only a finite number of values can be stored in processing systems. This results in a maximum and minimum value that can be stored and a finite distance between the quantization steps. This distance can be characterized by the Least Significant Bit (LSB). The quantization error is correlated with the original signal. Depending on the amplitude of the signal and the number of bits used for quantization this leads to audible distortion artifacts (VANDERKOOY and S. P. LIPSHITZ, 1984). If the signal amplitude can be assumed much larger than the LSB the quantization error can be assumed similar to Additive White Gaussian Noise (AWGN). In this case the quantization error has a mean value of zero and a standard deviation of $1/\sqrt{12}$ multiplied by the LSB (JIMENEZ, L. WANG, and Y. WANG, 1991).

For audio signal processing, a technique called *dithering* is usually applied to distribute these quantization errors over the frequency range. This technique is commonly applied in the field of digital image processing or printing as well. Due to the masking effect of our ears these artifacts become less audible. Dithering describes a method where noise is added to the original signal prior to quantization to break the correlation of the quantization error with the input signal. Depending on the amount and the spectral shape of noise added the SNR is potentially reduced but the perceived quality rises (S. LIPSHITZ, WANNAMAKER, and VANDERKOOY, 2004; VANDERKOOY and S. P. LIPSHITZ, 1984). The origin of the noise is not important, i.e. electrical noise that overlays the input signals might act as dithering noise as well. This quantization effect is usually not considered when measuring transfer functions and hence additional dithering is not applied. The three different Propability Density Functions (PDFs) mostly used for dithering are the Gaussian, rectangular and triangular distribution. See Section 2.4.2 for details and examples. The peak-to-peak amplitude of the dithering noise typically chosen is one LSB for the rectangular and two LSBs for the triangular distribution, called *AES17 dithering* in the following. The standard deviation of the Gaussian noise is often chosen as one LSB as in the *triangular AES17* dithering which is used in this thesis (AES17, 1998).

2.3. Parametric Transfer Function Models

The description of acoustical scenarios with a parametric model, where the parameters can be modeled with associated uncertainties, is essential for this work. Such a parametrized model is already used in the field of signal processing for transfer functions in the Laplace or discretized z-domain. In signal processing the modeling of transfer characteristics with poles and zeros is common and as it provides a compact, parametric description. A transfer function can be approximated or even analytically expressed by a rational function $H(s)$ in the Laplace domain as

$$H(s) = \frac{P_N(s)}{Q_M(s)} \qquad (2.3)$$

with the polynomial functions $P(s)$ and $Q(s)$ of degree N and M. The link between these abstract poles and zeros and especially their relation with the physics of a scenario to be modeled strongly depends on the field of application.

2.3.1. Poles, Zeros and Resonators

Poles and Zeros The roots of the numerator polynomial $P(s)$ are called *zeros* n_i and the roots of the denominator $Q(s)$ are called *poles* p_i. According to the fundamental theorem of algebra a complex polynomial of degree N has N roots. Hence, the transfer function can be written as a multiplication of these roots as

$$H(s) = \frac{\displaystyle\prod_{i=1}^{N} (s - n_i)}{\displaystyle\prod_{i=1}^{M} (s - p_i)}. \qquad (2.4)$$

The representation in Eq. 2.4 is called *pole zero representation*. Numerical problems in the determination of the roots usually increase with higher polynomial orders. The complex transfer function $H(s)$ is directly related to the impulse response $h(t)$. Hence, a symmetry in the distribution of poles and zeros according to the real axis can be observed. Poles and zeros that do not lie directly on the real axis can be associated with a counterpart being the conjugate complex. The rational transfer functions can be divided into two parts: Firstly, all-pole or Auto Regression (AR) models with $P(s) = 1$ directly related to Infinite Impulse Response (IIR) filters and secondly, all-zero or Moving Average (MA) models with $Q(s) = 1$ leading to Finite Impulse Response (FIR) (TOHYAMA and KOIKE, 1998).

Single Resonator   The real-valued eigenfrequency f_i, the real-valued damping constant σ_i and the amplitude c_i which acts generally as a complex scaling factor, characterize the resonator. The eigenfrequency and damping constant can be summarized in a variable called the *pole* p_i (TOHYAMA and R. H. LYON, 1989; TOHYAMA, R. LYON, and KOIKE, 1994)

$$p_{\pm i} = \sigma_i \pm j2\pi f_i \tag{2.5}$$

in the following which are symmetric regarding the damping axis. Except for poles with the eigenfrequency $f_i = 0$ they always appear in pairs for systems with a real-valued representation in the time domain. For acoustic or mechanical systems the physical link between the poles is provided by the acoustic or mechanical eigenfrequencies and the damping. The coefficient c_i is in general complex and can be written with real-valued α and β as

$$c_i = \alpha_i + j\beta_i. \tag{2.6}$$

The formulation of the transfer function H_{single} of a single resonator in the s-plane is given as

$$H_{\text{single},i}(s) = \frac{c_i}{s - p_i} + \frac{c_i^*}{s - p_i^*} \tag{2.7}$$

and the associated formulation in the time domain reads as

$$h_{\text{single},i}(t) = \mathcal{Re}\left(c_i \cdot e^{-t/\tau} \cdot e^{j2\pi f_i t} \right). \tag{2.8}$$

15

The expression for the pairs of poles in the frequency evaluated at $s = j\omega$ domain can be written as

$$H_i(\omega) = \frac{c_i}{j\omega - p_i} + \frac{c_i*}{j\omega - p_i*} = \frac{\alpha_i + j\beta_i}{j\omega - \delta_i - j\omega_i} + \frac{\alpha_i - j\beta_i}{j\omega - \delta_i + j\omega_i} = 2\frac{\alpha(j\omega - \delta_i) - \beta\omega_i}{-\omega^2 - 2j\omega\delta_i + \delta_i^2 + \omega_i^2} \tag{2.9}$$

where the latter term is similar to the term used for modes in mechanics or room acoustics. Such a single resonator was shown in Figure 2.3.

Multiple Resonators – Pole-Residue Model   In real-world scenarios usually more than one resonator occurs in the frequency range of interest. Hence, the combination of these single resonators is observed. This combination is also called *modal superposition* which is represented in the following equation by the sum:

$$H(s) = \sum_{i=1}^{N} H_{\text{single},i} = \sum_{i=1}^{N} \frac{c_i}{s - p_i} + \sum_{i=1}^{N} \frac{c_i^*}{s - p_i^*}. \tag{2.10}$$

Eq. 2.10 is the partial fraction decomposition or expansion of Eq. 2.4, i.e. these representations can be transformed into each other. The poles remain identical in this transformation but the coefficients c_i depend on the poles p_i and the zeros n_i of Eq. 2.4. In the context of this decomposition the coefficients c_i are often called *residues*. Hence, the representation in Eq. 2.10 is called *pole residue representation*.

2.3.2. Common Acoustic Poles and Zeros Approach

The Common Acoustic Poles and Zeros approach (CAPZ) takes into account that eigenfrequencies and their associated damping constants (poles), e.g., in room acoustics or in structure-borne scenarios remain constant over the entire system but only the coefficients of these poles change with source and receiver position (HANEDA, MAKINO, and KANEDA, 1994). This approach was also applied to characterize directional microphones, especially binaural receivers (HANEDA et al., 1999). Nevertheless, this approach is applicable for arbitrary acoustical systems as it is physically motivated by the formulation used in modal superposition. By considering this principal behavior when analyzing modes in acoustic systems the number of required modal parameters to represent an acoustic Multiple Input Multiple Output (MIMO) system can be reduced. A single set of common poles (eigenfrequencies and damping constants) is stored for the entire system and only the zeros are stored for different source or receiver positions separately. By using the relation between both representations, the CAPZ can be generalized as far as the coefficients or residues are depending on the specific position.

2.3.3. Rational Fit and Vector Fit

Vector Fitting is a robust macro-modeling tool for approximating frequency domain responses of complex physical structures (DESCHRIJVER, HAEGEMAN, and DHAENE, 2007). An open-source MATLAB routine realizing the iterative fitting for data in the frequency domain with various a-priori constraints to be set as optional parameters is available and described in detail along with the underlying algorithms in (GUSTAVSEN and SEMLYEN, 1999). This routine uses the pole-residue model. The fitting problem is linear and overdetermined, i.e., the number of poles has to be smaller than half the number of frequency bins available for the input. The algorithm is based on placing a set of complex starting poles with arbitrary distribution in the s-domain. Then the residues are identified by using the input data. The iteration is used to redefine the location of the poles. Noise has an impact on quality of the fit and hence low-noise measurement data is required (GUSTAVSEN and SEMLYEN, 1999).

In case a-priori knowledge about the system is available the starting pole configuration might be adapted to physically meaningful values. This covers the frequency interval between the resonances as well as limitations of the frequency range. In acoustics, the modal density is generally dependent on frequency. Hence, the starting poles should be distributed in a similar manner to achieve faster convergence and better fitting results. Since modal densities in acoustics usually rise over frequency the fitting can only be used up to maximum frequency f_{max}. The number of poles N_{poles} (positive and negative frequency axis) that could be fitted by the algorithm depend on the number of frequency bins available (including the negative frequency axis) and hence on the length of the impulse response T_{max} as

$$N_{\mathrm{poles}} = T_{\mathrm{max}} \cdot f_{\mathrm{max}} \tag{2.11}$$

where a useful number of poles is the rounded value towards the lower integer value instead.

2.4. Uncertainty Modeling Techniques

The term uncertainty is used differently in the literature. It might refer to *variation*, sometimes just *noise* or to (systematic) *errors*. Two different types of errors are distinguished—*systematic* and *random* errors. The most straight-forward approach to determine the uncertainty of a measurement is to measure a quantity with the exact same measurement setup several times. The resulting distribution of the value shows a range of possible solutions and a PDF. This approach cannot be used to determine systematic errors but only random errors.

2.4.1. Guide to the Expression of Uncertainty in Measurement

The Guide to the expression of Uncertainty in Measurement (GUM) is a document providing mathematical and theoretical background as well as a guideline to determine the measurement uncertainty in a unified manner (JCGM 100, 1995) and it is increasingly referred to from standards claiming for the assessment of the measurement uncertainties of the defined quantities. A *measurand* is defined as the quantity to be measured and also called *output quantity* Y. This output quantity is dependent on various factors, e.g., meteorological conditions. Sometimes the output quantity cannot be directly measured but is determined by further calculation of multiple measured quantities. All these factors are called input quantities X_i. In theory, a mathematical function defines the relationship between all important N input quantities with the output quantity as: $Y = f_M(X_1, X_2, \ldots, X_N)$. Only for a small percentage of applications this *model function* f_M can be expressed analytically. However, this model function and the determination of the most important input quantities is the central element in uncertainty modeling.

The GUM distinguishes between *type A* and *type B* uncertainties and not between systematic and random. *Type A* are characterized by statistical observation and *Type B* uncertainties are are based on *other scientific knowledge*. The latter includes simple estimates from experts.

The official guideline or scheme to obtain the correct uncertainty values is a seven step procedure:

1. Collect information on the measurand Y and the input quantities X_i;
2. Find a suitable model function f_M;

3. Evaluation of the input quantities according to "type A" or "type B" uncertainties;

4. Calculation of the result in terms of mean value "y" and range of uncertainty $u(y)$;

5. Obtaining the complete measurement results as $y \pm U$;

6. Calculation of the measurement uncertainty budget.

This thesis focuses on the principal tools to model the uncertainty contributions and the application in the field of acoustics. It does not provide the complete evaluation and assessment of the range of uncertainty according to the GUM.

2.4.2. Distribution and Probability Density Functions

The most common PDFs are the rectangular, triangular and Gaussian or normal distribution that are also used in this thesis. All distributions are completely characterized by their type, a mean value μ and a parameter specifying the spread (BRONSHTEIN et al., 2005).

The *rectangular* or uniform or equal distribution is defined as

$$\rho_{\text{rect}}(x) = \frac{1}{2b} \quad \text{for} \quad \mu - b < x < \mu + b \quad \text{and} \quad 0 \quad \text{otherwise} \tag{2.12}$$

with b specifying the range of the distribution.

The resulting PDF of a sum of two variables with uniform distributions is called a symmetric *triangular* distribution. It can also be expressed as a convolution of the two rectangular PDFs. It is defined as

$$\rho_{\text{tri}}(x) = \frac{1}{b^2} \left(|x - \mu| + b \right) \quad \text{for} \quad \mu - b < x < \mu + b \quad \text{and} \quad 0 \quad \text{otherwise.} \tag{2.13}$$

Finally, the Gaussian or normal distribution can be achieved when an infinite number of variables is superposed. It is defined as

$$\rho_{\text{gauss}} = \frac{1}{\sigma \sqrt{2\pi}} \mathrm{e}^{-\frac{1}{2} \frac{x-\mu}{\sigma}^2} \tag{2.14}$$

with σ being the standard deviation. This distribution has the advantage that the sum of two variables results in the same type of distribution and the mean value and standard deviation of the result can directly be expressed. Hence, variables are preferably associated with normal distribution due to simplicity, especially in practical applications of the GUM.

2.4.3. Monte-Carlo Simulations

Monte-Carlo Simulations are applied in uncertainty modeling when the model function f_M cannot be expressed analytically or the PDFs of the input quantities do not allow a direct determination of the PDF of the result (COX and SIEBERT, 2006). In these simulations, the input quantities are varied randomly according to the PDF of the input quantities and the output is calculated numerically. For a sufficiently large number of simulations the mean value and the PDF of the output quantity can be analyzed. For computationally complex models, e.g., including FEM simulations this method requires long computation times. This method becomes very efficient with analytic models or models with low computational complexity as used in this thesis. It is important to mention that although the transfer function might be expressed analytically the derived parameter can not always be expressed analytically. Hence the model function remains unknown in Monte Carlo (MC) simulations are used to estimate this function.

3

Measurement of Acoustic Systems

This chapter describes the measurement procedure of transfer functions of acoustic systems in detail. The state-of-the-art measurement apparatus and the measurement chain connecting the apparatus is introduced to allow the determination of absolutely calibrated Transfer Functions (TFs). Common challenges in the determination of the transfer characteristics, violations against the Linear Time-Invariant (LTI) assumptions and practical aspects—typically not considered in theoretical textbooks—of such measurements are addressed. As post-processing of measured transfer functions can yield significant improvement in terms of quality and reducing artifacts, also influencing the measurement uncertainty, these methods are explained. Contributions to the characterization of nonlinear systems are presented to finally allow the estimation of the measurement uncertainty with nonlinear elements in the measurement chain. These effects of the measurement procedure on the measurement uncertainty of the transfer function are finally discussed for a practical example from the field of room acoustics.

3.1. Characterization of LTI Systems

Modeling and measuring an LTI systems by means of a black box approach is a common task in industry and research and typically illustrated similar to Figure 3.1. The correct determination of the transfer function or the corresponding Impulse Response (IR) is the major goal. A detailed overview of transfer function measurement for electro-acoustic systems is given by MÜLLER AND MASSARANI (MÜLLER and MASSARANI, 2001). However, most principles are directly applicable to acoustic systems in general. A simplified measurement chain with signal excitation $S(f)$, Digital to Analog Converter (DAC), the Device Under Test (DUT), the Analog to Digital Converter (ADC) and required deconvolution by

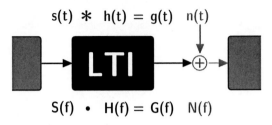

Figure 3.1.: Black box modeling approach of an LTI system with description in either time and frequency domain: Input signal $s(t)$ or $S(f)$, system output $g(t)$ or $G(f)$, system impulse response $h(t)$ or transfer function $H(f)$ and measurement noise $n(t)$ or $N(f)$.

multiplication with the inverse of the complex excitation spectrum is illustrated exemplarily in Figure 3.2. Generally, the calculation of the transfer function by using the system output and a known input signal is called *correlational measurement*. The determination with unknown input signals is denoted as *blind system estimation* and is not considered in this work[1].

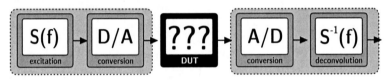

Figure 3.2.: Simplified digital measurement chain to determine the transfer function of the DUT with an ex n signal

3.1.1. Excitation Signals and Deconvolution

While the determination of the impulse response with a Dirac impulse at the input of the system is a simple and straightforward approach, it is most problematic in terms of Signal-to-Noise Ratio (SNR). Theoretically, any arbitrary excitation signal could be used, e.g., music, speech, noise or technical signals, that contains sufficient energy for a broad frequency range. Typical technical excitation signals are pure tones, noise, Maximum Length Sequence (MLS), or sweeps. Although pure tones are often found they are not broad-band and therefore not considered further in this thesis. MLS are sometimes used with the deconvolution techniques described below, but then these signals are practically treated as broad-band noise.

---

[1] Parts of this section have already been published in (DIETRICH, LIEVENS, and PAUL, 2009) and (DIETRICH, MASIERO, and VORLÄNDER, 2013).

The proper post-processing for MLS uses the *Hadamard transform* (NELSON and FREDMAN, 1970).

Signal Types—A brief Overview

Pure sine or cosine tones have advantages regarding measurement equipment or the measurement of nonlinear distortion. Total Harmonic Distortion (THD) or the maximum sound pressure level (maxSPL) of pro-audio loudspeakers is commonly measured with pure tones. THD is defined as the ratio of the magnitude of the harmonics g_k of order k and the magnitude of the fundamental g_1 as (BALLOU, 1987)

$$\text{THD} = \frac{\sqrt{\sum_{k=2}^{\infty} g_k^2}}{g_1}. \tag{3.1}$$

If spectra of the fundamentals and the harmonics $G_k(f)$ are available the THD spectrum can be expressed as

$$\text{THD}(f) = \frac{\sqrt{\sum_{k=2}^{\infty} |G_k(k \cdot f)|^2}}{|G_1(f)|}. \tag{3.2}$$

E.g., for vibration isolators in structure-borne acoustics pure tones are still used. For a sufficient frequency resolution several measurement with different frequencies have to be performed. As this type of signal is narrow-band it is not directly applicable for broad-band correlational measurements. The SNR achievements are mostly sufficient as the focus on a specific frequency with a system in steady-state can exclude the noise at all other frequencies.

Random noise as a measurement signal is still used in many applications nowadays, mainly due to old measurement systems that are limited in their signal processing methods. Different colorations of random noise—sometimes still generated by analog electrical circuits using the electrical noise in resistors—can be found distinguished by their spectral shape. The spectral shape considers the power over frequency. These noise signals are commonly used as non deterministic signals. White noise is per definition broad-band and covers theoretically all frequencies. As in digital signal processing all signals are low-passed the term *all frequencies* means up to the NYQUIST frequency instead. Important shapes are *white* (flat, 0 ᵈᴮ/oct.) *pink* (falling, -3ᵈᴮ/oct., -10 ᵈᴮ/decade), *brown* (-6 ᵈᴮ/oct.), *blue* (rising, $+3$ ᵈᴮ/oct.).

In the following a signal is discussed that should be preferred over noise as it can have similar spectral and tempo-spectral characteristics and is deterministic instead. The MLS method is efficient in terms of computational complexity by using the fast implementation of the Hadamard transform—the Fast Hadamard Transform (FHT)—as proposed in (BORISH and ANGELL, 1983). With increasing computer power and memory this advantage is not longer important (MÜLLER and MASSARANI, 2001). Due to the fact that the correlation with the MLS can be realized by only one FHT and some shift operations no further Fast Fourier Transform (FFT) and no deconvolution is required. This technique works directly in time domain and the result is the impulse response. The classical MLS is broad-band.

Sweeps—synonyms used in literature are *chirps* or *swept-sines*—continuously drive through a frequency range with arbitrary slew rates. Important types are the *linear* and *exponential*—sometimes called logarithmic sweep—sweep with a linear and an exponential relationship between frequency over time. Linear sweeps can be generated as broad-band signals. Exponential sweeps have the theoretical drawback to start at a finite lower frequency value. For practical considerations, sweeps can be easily designed to cover a limited frequency range but without having energy zeros in this frequency range. This thesis concentrates on sweeps as excitation signals only due to their advantageous described in the following.

The transfer function $H(f)$ can be simply expressed as the ratio of the output signal $G(f)$ by the input signal $S(f)$ in frequency domain. The impulse response $h(t)$ follows directly as the inverse Fourier transform of $H(f)$. This method is called *deconvolution* analog to the well known *convolution* where a multiplication of two spectra is used instead. Despite the time domain methods, that use direct expression of the inverse filter in time domain as, e.g., possible for specific sweeps, this deconvolution can be generally realized more efficiently in frequency domain by transforming input and output signal with the Fourier trace($S(f)$ and $G(f)$)—and later the FFT in the digital domain—and processing a spectral division to obtain the complex transfer function of the system

$$H(f) = \frac{G(f)}{S(f)} \quad \text{and} \quad S(f) \neq 0. \tag{3.3}$$

Generally, numerical problems arise for frequencies where the absolute values of $S(f)$ become very small. Hence, the signal should theoretically be broad-band. A more relaxed requirement claims sufficient energy in the frequency range of interest, e.g., the audible frequency range from 20 Hz to 20 kHz. The output of the system is always superposed by measurement noise ($n(t)$ or $N(f)$) as illustrated in Figure 3.1. This noise will be amplified by the deconvolution leading to errors in the impulse response.

Cyclic and Linear Deconvolution

In the digital domain the Discrete Fourier Transform is used and one approach for the deconvolution process is by simple dividing the frequency bins of the two spectra separately. The overall number of frequency bins remains the same and the length of the corresponding impulse response is equal to the length of the excitation signal. This is called *cyclic deconvolution* since the behavior in time domain can be considered as cyclic shifts. E.g., the cyclic deconvolution of a Dirac impulse at 0 s with a Dirac at 2 s and an overall signal duration of 10 s results in a Dirac at 10 s − 2 s = 8 s due to this cyclic shift. This scenario is not causal, but this problem occurs in real life when the system output contains noise that is not correlated with excitation or if the system shows nonlinear behavior and if it is measured with sweeps as discussed in Section 3.3.3. The same final response would occur if an impulse at 10 seconds was deconvolved with the same impulse at two seconds. This is then a causal scenario, but it cannot be distinguished from the final result alone if causal or non-causal behavior leads to this, e.g., this Dirac in the example. Zero padding of the discrete time signals s and g in the end in time domain and extending the length of these signals to twice the original length prior to division can therefore overcome this effect. This approximates the *linear deconvolution*.

Especially for linear and exponential sweeps the *inverse sweep* or the matched filter of the time streched pulse can also be calculated directly in time domain as, e.g., formulated in (NOVAK, 2009; SUZUKI et al., 1995). This inverse sweep is then convolved with $g(t)$ and no further regularization to account for the band limitation is required and it already has the same frequency limits as the original sweep. However, the band limitation is hard and approximately a *rect* function in frequency domain and hence the resulting impulse response suffers from severe pre-ringing and post-ringing similar to its inverse Fourier transform, which is the *sinc* function.

For the non-causal scenario the result changes to an impulse at $20\,\text{s} - 2\,\text{s} = 18\,\text{s}$ and the result for the causal scenario remains untouched. With this technique an easy separation between causal and non-causal behavior is realized. All components appearing before half the duration of the impulse response are causal and all components beyond this point are non-causal. Non correlated signal components, e.g., noise, are distributed over the entire impulse response. The linear deconvolution technique increases the computational effort for the Discrete Fourier Transform (DFT). However, the noise floor of the impulse response is not constant over time as frequency limits of the noise vary over time according to the sweep rate. It reaches its minimum at half the length of the impulse response. Figure 3.3 illustrates the difference between both deconvolution techniques with a measurement in a room. The variation of the background noise over time can be observed directly in the spectrograms. The peaks in the end of the impulse response are due to nonlinearities and will be explained in Section 3.3.

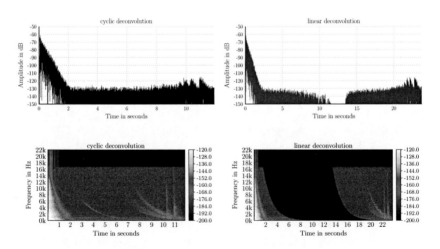

Figure 3.3.: Difference between cyclic (left) and linear (right) deconvolution technique in a measured room impulse response (auditorium) including weak nonlinearities of the loudspeaker.

Zero-phase Regularization

To avoid the division by small values in the spectrum FARINA (FARINA, 2007) introduced a regularization method already used by KIRKEBY and many others in different contexts. The regularized inverse reads as follows

$$S_{\text{inv,reg}}(f) = \frac{S^*(f)}{S^*(f)S(f) + \varepsilon(f)} \qquad (3.4)$$

where $S^*(f)$ denotes the complex conjugate of $S(f)$ and $\varepsilon(f)$ is a real but frequency dependent regularization parameter. The regularized transfer function follows accordingly as

$$H_{\text{reg}}(f) = \frac{G(f)}{S(f)} \frac{1}{1 + \varepsilon(f)/|S(f)|^2} = \frac{G(f)}{S(f)} \cdot A_{\text{reg}}(f), \qquad (3.5)$$

where $A_{\text{reg}}(f)$ describes the influence of the regularization as a filter. Since the input signal is band-limited it is reasonable to increase the regularization parameter below f_1 the lowest frequency and above f_2 the highest frequency of interest. The resulting transfer function is also band-limited and is less influenced by measurement noise. The example shown in Figure 3.3 used $f_1 = 50\,\text{Hz}$ and $f_2 = 17\,\text{kHz}$. The phase of $A_{\text{reg}}(f)$ is zero for all frequencies but this filter is band-limited. Hence, the equivalent impulse response of $A_{\text{reg}}(f)$ is symmetric regarding the time axis, i.e., the impulse response has a non-causal part depending on these band-limiting frequencies f_1 and f_2.

Minimum-phase Regularization

To avoid non-causality, $A_{\text{reg}}(f)$ can be factorized into a minimum-phase (MP) regularization filter and a remaining non-causal all-pass (AP) filter (TOHYAMA and KOIKE, 1998) (BOUCHARD, NORCROSS, and SOULODRE, 2006):

$$A_{\text{reg}}(f) = A_{\text{reg,MP}} \cdot A_{\text{reg,AP}}(f). \qquad (3.6)$$

By using $A_{\text{reg,MP}}$ in the deconvolution process the obtained fundamental impulse responses are always causal. Time windowing can be applied in a consecutive step to suppress unwanted signal components as described in Section 3.6. As a last step of data post-processing the all-pass component can be applied to compensate for the phase error in the pass-band yielding non-causal impulse responses again. This method has advantages when the arrival time of an impulse is known and the time window should be applied directly at these pre-calculated arrival times. In the context of multiple excitation signals the measured and deconvolved signal have to be split into the impulse responses of separate systems. Non-causal impulse responses are problematic as the signal can spread to the left on the time axis and possibly interfere with the end of impulse response of the previous system in case several systems are measured in a parallel manner (DIETRICH, MASIERO, and VORLÄNDER, 2013; MAJDAK, BALAZS, and LABACK, 2007). Hence, this splitted post-processing as described above should be preferred.

Analytical Formulation of Sweeps

The following analytical formulation of the sweep is based on (HUSZTY and SAKAMOTO, 2010; MÜLLER and MASSARANI, 2001; NOVAK, 2009). A sweep signal is defined in time domain by

$$s(t) = \sin(\phi_{\text{inst}}(t) + \phi_0) \tag{3.7}$$

with its time varying *instantaneous* phase component $\phi_{\text{inst}}(t)$ and the starting phase ϕ_0. It is convenient to choose the starting phase $\phi_0 = 0$ as this results in a smooth start of the signal without a discontinuity in the beginning. This is important as the signals for practical measurements have a finite length and therefore a finite frequency range. The instantaneous frequency $f_{\text{inst}}(t)$ over time is defined as (NOVAK, 2009)

$$f_{\text{inst}}(t) = \frac{1}{2\pi} \frac{\mathrm{d}\phi_{\text{inst}}(t)}{\mathrm{d}t}. \tag{3.8}$$

Commonly this instantaneous frequency is chosen either as a linear or as an exponential function over time and the instantaneous phase is not directly given. An exponential sweep starts at its lowest frequency f_1 and increases the frequency to its highest frequency f_2 in an exponential manner over time defined by the *sweep rate*

$$r_{\text{sw}} = \frac{\log_2\left(f_2/f_1\right)}{\tau_{\text{sw}}} \text{ octaves/s} \tag{3.9}$$

with the time τ_{sw} between those frequencies. As the following formulation uses the basis e instead of the basis 2 the rise time constant L_{sw} is introduced as

$$L_{\text{sw}} = \frac{\log_2(\mathrm{e})}{r_{\text{sw}}} = \frac{\tau_{\text{sw}}}{\ln\left(\frac{f_2}{f_1}\right)}. \tag{3.10}$$

The instantaneous frequency is therefore given as

$$f_{\mathrm{inst}}(t) = f_1 \cdot \mathrm{e}^{t/L_{\mathrm{sw}}} \quad \text{and} \quad t \in [0, T] \tag{3.11}$$

and is zero otherwise. The instantaneous phase of the sweep can now be obtained due to Eq. 3.8 by integration of Eq. 3.11 as

$$\phi_{\mathrm{inst}}(t) = 2\pi \int_0^t f_{\mathrm{inst}}(\tau)\, \mathrm{d}\tau = 2\pi f_1 L_{\mathrm{sw}} \left(\mathrm{e}^{t/L_{\mathrm{sw}}} - 1 \right) \tag{3.12}$$

and the detailed formulation of the exponential sweep in time domain follows:

$$s(t) = \sin\left(\underbrace{2\pi f_1 \cdot L_{\mathrm{sw}} \cdot \mathrm{e}^{t/L_{\mathrm{sw}}}}_{\mathrm{time-dependent}} + \underbrace{\phi_0 - 2\pi f_1 \cdot L_{\mathrm{sw}}}_{\mathrm{const.}} \right). \tag{3.13}$$

The measurement with an exponential sweep requires a certain time to allow the system to decay after the sweep has stopped. This time is introduced as a *stop margin* τ_{st} which is the duration when silence is played back while still recording the decay of the system.

3.1.2. Noise Influences

Every measurement contains noise that overlays the measurement signal. Each element of the measurement chain might introduce noise with different characteristics. The spectral shape of the noises differ as, e.g., acoustic background noise in a measurement laboratory with a frequently used street in the vicinity has a typical spectrum different from just simple white or pink noise typically assumed to be added in electrical units. Hums occur mainly due to ground loop or unsymmetrical resistance in the signal and the ground cables and connections. This is explained in detail with a novel approach in finding and eliminating ground loop problems by WHITLOCK in (BALLOU, 1987).

The fundamental frequency of the hum is commonly the frequency of the electricity network and therefore typically either $50\,\mathrm{Hz}$ or $60\,\mathrm{Hz}$. Harmonics of this fundamental frequency appear with different amplitudes depending on the measurement equipment and the running electric devices connected to the same network. The hum appears as a sharp spike in frequency domain and might be analyzed as, e.g., a resonance with low damping if not classified as noise. Hence, hums should be detected and preferably solved by rewiring the measurement setup. In transfer function measurements the hums signal is not correlated with the measurement signal but as it is mono-frequent it will not be rejected by, e.g., a sweep measurement. Notch filters with high quality might be used to eliminate these frequencies from the final measurement result but they always introduce a modification of the real measurement result as well. In some countries extra information or control signals are sent over the electricity network. In Germany, e.g., a clock synchronization signal is sent in fixed time intervals. Hence measurements should not overlap with these time slots. Both effects are not further considered in the simulations in this thesis.

The quantity related to the quality of the measurement is the SNR over frequency. Methods are available to pre-emphasize the measurement signal to equalize and improve the SNR, e.g., in (MÜLLER and MASSARANI, 2001). In particular, for the measurement of impulse responses the SNR can be improved with an increase in measurement time if and only if the system is time-invariant. Hence, for typical acoustic systems the measurement noise is not the main focus in terms of measurement uncertainty. But the influence of the noise and its spectral shape can be easily analyzed by superposition with the ideal impulse response if a valid model of the measurement scenario is available.

Although the background or measurement noise might be constant in level over the time of a measurement, the noise level in the measured impulse response might vary over time in sweep measurements, e.g., in case linear deconvolution is used (c.f. Figure 3.3). Impulsive acoustic noise might occur, which is transient and not constant over time. This behavior can also be observed in the previously shown spectrograms in Figure 3.3 as inverse sweeps. The impulse smears backwards over time starting at the time where it actually appeared. Furthermore, for the cyclic deconvolution virtual impulsive noise might occur at the beginning and the end of measurement. This can be observed in the spectrograms, where both parts shown in the linear deconvolution adjoin in the spectrogram for the cyclic deconvolution. These effects only occur due to a necessary limitation of the measurement duration to a finite length.

In order maximize the SNR and hence to obtain a stable impulse response of the DUT the level or the amplitude of the measurement signal has to be maximized or the measurement duration has to be increased. An extension of the measurement duration by a factor N results in an increase of the peak SNR of

$$\Delta\mathrm{SNR} = 3 \cdot \log_2\left(N\right) \mathrm{dB}. \tag{3.14}$$

This equation can also be used to approximate the gain in SNR by using extended signals with $N_2/N_1 \gg 1$ compared to impulses that have $N_1 = 1$. Both approaches become problematic if the system is not fully LTI and hence these topics are closely related. For nonlinear systems the increase in level might introduce more distortion as explained in Section 3.3. For time-variant systems the extension of the measurement duration might lead to artifacts as, e.g., shown for an airborne sound example in Section 4.5.3.

3.2. The Acoustic Measurement Chain

A typical measurement chain for airborne acoustic measurements is given in Figure 3.4 starting with the software signal output and reaching the software's signal input again. Only a few measurements require an absolutely calibrated measurement chain but an relative calibration between input channels or subsequent measurements instead. But for the sake of completeness and to analyze the behavior in a general way the transfer functions or transfer factors of each element are included using physical units. The *stars* in the figure correspond to the points where the signal can be measured. This simple diagram shows that every measurement chain can be fully calibrated step by step or at least with a quick electrical reference measurement and, e.g., a microphone calibration at a single frequency if the blocks in the input measurement chain can be assumed to have a flat frequency response. The DUT in acoustics is either the actuator, the sound transmitting element, the sensor or a combination of these. Hence, an electrical reference by simply connecting the electrical input and output and using this measurement result as a calibration is not sufficient in general. But this approach could minimize calibration errors. Any errors or uncertainties during the calibration procedure remain as uncertainties of the Frequency Response Function (FRF) of the DUT mainly as frequency independent amplitude or simple constant or linear phase errors.

In particular the following attributes have to be analyzed as they might be main sources of measurement uncertainties:

- The LTI frequency and phase response or impulse response;
- Nonlinear behavior or valid amplitude range of linearity;
- Measurement noise;
- Drift over temperature, pressure, time, etc. (Chapter 4 and Appendix A.4);

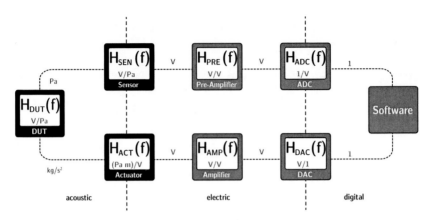

Figure 3.4.: Typical measurement chain in a block diagram for airborne acoustic measurements including physical quantities for absolutely calibrated measurements of transfer functions.

- Spatial radiation or sensitivity pattern (Chapter 4);
- Position of actuator or sensor relative to the DUT (Chapter 4 and Chapter 5);
- Cross-talk from other measurement paths (Chapter 5).

The ADC has in general a non-flat frequency response providing an anti-aliasing filter with a cut-off frequency around half the sampling rate and a high-pass in the range of approximately 1 Hz up to 20 Hz to suppress DC offsets in acoustical measurements [2]. This influence is assumed to be negligible. ADCs and DACs show in fact a nonlinear behavior as introduced in Section 2.2.5. The effects on the measurement will be studied in Section 3.5.3. Both are very stable over time, e.g. investigated in (POLLOW et al., 2011) and might introduce some extra noise but mainly around the level of the quantization noise. Power amplifiers and pre-amplifiers are nowadays powerful in terms of linearity, flat frequency response and noise suppression. But especially for power amplifiers an effect has to be pointed out that might introduce large uncertainties. High level input signals result in a high power output. This sudden change in power results in a discharge of the capacities in the power supply leads to a temporarily lower supply voltage. Hence, the linear range might be temporarily decreased possibly leading to distortion or the sensitivity of the amplifier is changed over time as described by the Common Mode Rejection Ratio (CMRR). Therefore, the

---

[2]Measurement equipment for structure-borne measurements sometimes has no high-pass filter.

power amplifier is possibly time-variant and nonlinear (GOERTZ, 2012) but not considered further in this thesis.

As discussed in many publications the loudspeaker remains the weakest part, e.g., (MÜLLER, 1999). It features a frequency dependent radiation pattern, a corrugated frequency response and nonlinear behavior. The frequency response can theoretically be compensated only if the source has the same frequency response for all radiation angles that contribute to the measured transfer function in an airborne acoustic setup without using further spatial processing. For a room acoustic measurement this was only possible if the source had an omnidirectional radiation pattern. This problem will be addressed in Section 4.1.

Measurement microphones can be characterized by their frequency dependent sensitivity which can be fairly flat for almost the entire audible frequency range. This sensitivity is in general dependent on the angle of the incident sound wave but can be very close to the omni-directional pattern compared to the one of typical measurement loudspeakers. The range of levels a realistic microphone can handle is limited to both ends. At low levels its inherent noise level limits the resolution and at high sound pressure levels the microphone becomes nonlinear.

Microphone pre-amplifiers mostly feature a high-pass characteristic far below the lowest frequencies of interest in acoustics, e.g., 20 Hz. However, the frequency response in the interesting range is fairly flat and they operate in mainly a linear range for typical levels of microphone signals. As these values are reported in the data sheets it is assumed that a suitable microphone is chosen for the sound pressure levels to be measured.

In order to study the principal influence of the elements of a typical measurement chain, a noise source before and after the DUT, a (frequency-dependent) nonlinear model representing the loudspeaker and the quantization introduced by ADC and DAC is used.

3.3. Nonlinear Systems

As the elements of the measurement chain are potentially nonlinear, loudspeakers have to be especially considered as such if the driving amplitude is high. But also ADC, DAC, amplifiers and sometimes DUTs in structure-borne sound (e.g., mechanical structures excited by high forces leading to high elongations) show nonlinear behavior. This section summarizes theoretical aspects from literature to model such systems. The influence of the measurement signal and the amplitude is discussed. An improved technique to determine nonlinear coefficients is introduced and finally a method to estimate the uncertainty caused by nonlinearities is presented.

An overview on the effects of nonlinearities in the measurement of acoustic transfer functions is given in (TORRAS-ROSELL and JACOBSEN, 2010). A detailed study of nonlinearities in acoustic is given in (NOVAK, 2009). The following sections will make use of his work in modeling and measurement of nonlinear systems and add improvements to the procedures. The reader is pointed to two shorter papers summarizing his approach (NOVAK et al., 2010; NOVAK, SIMON, and LOTTON, 2010).

3.3.1. Basic Nonlinear Model

Nonlinearities are problematic since the measured impulse response of a such a nonlinear system does not comply to the signal processing rules for LTI systems and hence produces artifacts which cannot be directly explained with these rules. A motivational example of this behavior already analyzed by various acoustic researchers in the past is given for a nonlinear transfer characteristic with a polynomial approach of order k without any frequency dependence and without a DC offset factor. The system output $g(t)$ writes as

$$g(t) = \sum_{i=0}^{k} a_i s^i(t) \tag{3.15}$$

with the polynomial coefficients a_i and the input signal $s(t)$ and can be drawn in a block diagram as in Figure 3.5, where x stands for the input signal of the block.

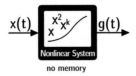

Figure 3.5.: Nonlinear block described by a polynomial with coefficients up to an order k without memory and without dependence on frequency.

A visualization of the nonlinear behavior is given by the *input-output-diagram* in Figure 3.6 for a polynomial of order 4 that reads $y = x + 0.1x^2 + x^3 + 0.1x^4$. The effects on the impulse response obtained by the exponential sweep, linear

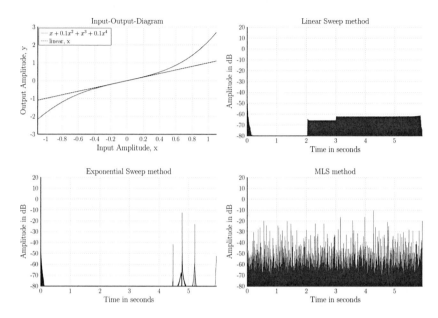

Figure 3.6.: Input-output-diagram of a nonlinear system with a polynomial approach ($H_a(f) = H_b(f) = 1$) and simulated impulse response of basic nonlinear system obtained with the linear sweep, exponential sweep and MLS method with a level of $0\,$dBFS each.

sweep and the MLS method with a driving amplitude of $0\,$dBFS[3] are shown in Figure 3.6. The linear transfer characteristic is a simple Dirac impulse, but the higher order nonlinearities add artifacts which are observable in the impulse responses. In case of the linear sweep method, harmonics are distributed over the time axis in the impulse response observed as backwards running sweeps with different rates, whereas they appear as sharp impulses in the exponential sweep method.

With the MLS method the nonlinear artifacts appear as spikes all over the impulse response. Each MLS has its own specific nonlinear spike pattern. Within one order of the MLS there exist multiple MLS possible. Averaging measurements

---

[3]FS stands for full scale. This is a value in dB smaller or equal to zero.

with these different MLS in order can significantly reduce the spikes. These artifacts in the impulse responses of MLS based measurements of weakly nonlinear systems have already been addressed in (RIFE and VANDERKOOY, 1989) and later in, e.g., (DUNN and HAWKSFORD, 1993; VANDERKOOY, 1994). However, the modeling of the nonlinear systems and the terminology has been refined in the last decade.

When it comes to nonlinear systems the *crest factor* or *peak-to-average ratio* becomes important as the maximum amplitude in time domain has to lie below a certain limit depending on the system to avoid significant nonlinear behavior. The crest factor of an ideal MLS is due to its similarity to a square wave form exactly 0 dB. Ideal linear or exponential sweeps have a crest factor of 3 dB similar to the sine wave form. This shows that MLS might be in average 3 dB louder than an sweep resulting in the same amount of increased SNR. However, a simple phase shift or an all-pass filter applied to the MLS signal prior to arriving at the nonlinear component in the measurement chain has completely changes this behavior and the advantage vanishes. The ideal Propability Density Function (PDF) of an MLS consisting of two Diracs at values ± 1 is transformed to a Gaussian PDF resulting in an average crest factor of approximately 11 dB (RIFE and VANDERKOOY, 1989). This would lead to a theoretical advantage of 8 dB in SNR for sweeps measurements. This explains the higher level of the artifacts in the simple example in Figure 3.6 for the MLS.

As the crest factor of the sweep is also dependent on the spectral shape these theoretical limits are generally not applicable in practical measurements unless the sweep signal is pre-emphasized in such a way that flattens the time envelope of the sweep at the entrance of the nonlinear component, i.e., compensating $H_a(f)$.

It is worth mentioning at this point, that the perfect Dirac delta of the linear part would result in a magnitude of 0 dB at the beginning of the impulse response. All measurement methods applied to the nonlinear system show a higher magnitude as the higher order polynomial parts interfere even at the beginning of the impulse response with the linear part. The frequency range for the sweeps has been chosen to 100 Hz up to the Nyquist frequency of 22050 Hz resulting in blurred impulses. Cyclic deconvolution is used for the sweeps. The MLS was directly used broad-band with proper correlation using the FHT. The length of the excitation signal equals the length of the impulse response as given in the plots.

Due to the advantageous in realistic crest factors and hence SNR and the often stated claim that sweep signals are capable of suppressing nonlinear distortion, sweep signals are solely investigated in the following.

The particular problem in context of nonlinear system behavior faced by measurement engineers can be simply formulated: How well does the measured impulse response capture the behavior of the DUT, i.e., its linear impulse response.

3.3.2. Wiener-Hammerstein Model

NOVAK has presented a method for the characterization of nonlinear systems using noise and exponential sweep signals. As the exponential sweep allows a less complex modeling and measurement approach and enables the measurement engineer to monitor nonlinearities by simply looking at the measured impulse response, only this class of sweeps is further considered. Detailed models for the description of electrodynamic loudspeakers have been developed by KLIPPEL, e.g., in (KLIPPEL, 1992). These models are mainly valid for low frequencies and a more generalized approach to assess nonlinear behavior of virtually any element in the measurement chain is used in this thesis. Furthermore, the model presented in the following can be directly used to analyze the uncertainties.

The basic *Wiener* and *Hammerstein* models are given in Figure 3.7 consisting of only a single static nonlinear without memory and a single dynamic linear time-invariant block. The block with memory is characterized by its transfer function $H_a(f)$ before and with $H_b(f)$ the transfer function after the nonlinearity. To model a complex system with frequency dependence, e.g., a loudspeaker, a combination of both—the *Wiener-Hammerstein* model—is required. In that case $H_a(f)$ would describe the transfer function between voltage and membrane velocity, the nonlinear block would restrict the membrane movement and especially its elongation and $H_b(f)$ would represent the radiation of the membrane including the enclosure into the free-field.

As an example for a Wiener-Hammerstein model a peak filter is used for $H_a(f)$. It is set to center frequency of 1 kHz with a quality factor of $Q = 10$ and a gain of $+10\,\mathrm{dB}$. The spectrum for the Hammerstein part is set to be the inverse as $H_b(f) = 1/H_a(f)$. For a linear polynomial the transfer function of this system is a flat spectrum with a magnitude $0\,\mathrm{dB}$ and a phase of $0°$. Hence, this is the *linear response*. The peak filters are shown in Figure 3.8. The polynomial is set to $x + 0.1x^2 + 0.1x^3 + 0.1x^4$ in order to obtain distortion in the order of magnitude

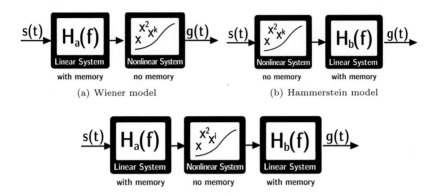

(c) Wiener-Hammerstein model

Figure 3.7.: Simple *Wiener* model as serial connection of dynamic linear block followed by static nonlinear block (left) and simple *Hammerstein* model as serial connection of static nonlinear block followed by dynamic linear block (right) and combined *Wiener-Hammerstein* model (bottom).

of the linear part. The simulation is conducted with an exponential sweep with a frequency range of 100 Hz to 20 kHz with a sweep length of approx. 12 s at a level of 0 dBFS. The extracted spectra of the harmonics are depicted in Figure 3.8 (right). As can be seen, the fundamental (indicated as harmonic 1) does not show the behavior as expected for the linear response. It shows a residual peak around 1 kHz due to signal energy that is transferred by the term x^3 to the fundamental as later explained in Section 3.3.5. The harmonics all contain a dip at 1 kHz due to $H_b(f)$. Furthermore, each harmonic k shows a peak at $f = k \cdot 1$ kHz. The high pass observed is caused by the lower cut-off frequency of the sweep and the regularization. This cut-off is k times higher for the harmonics. Although only a few parameters are used for this simulation it can be seen that the interpretation of the resulting impulse response is more complex than for linear systems.

The fundamental impulse is not equivalent to the linear response if the system is nonlinear and its spectral components will change depending on the excitation signal.

The polynomial Wiener and Hammerstein model is a generalization of the former simple models and depicted in Figure 3.9. The nonlinear blocks are narrowed to describe only the transfer characteristics of a single polynomial component $()^k$ of order k. The associated dynamic linear blocks weight the contribution of

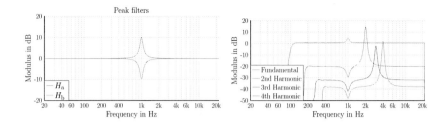

Figure 3.8.: Peak filters (left) $H_a(f)$ and $H_b(f)$ used in the Wiener-Hammerstein model and frequency dependent harmonics (right) observed with a polynomial $x + 0.1x^2 + 0.1x^3 + 0.1x^4$ and $0\,\text{dBFS}$.

the nonlinear component dependent on frequency for each order separately with $H_{b,k}$ or pre-emphasize the signal at the input of the nonlinearity by $H_{a,k}(f)$. For more details and background it is referred to (NOVAK, 2009) and (JANCZAK, 2004). In the following the basic models will be used for the examples.

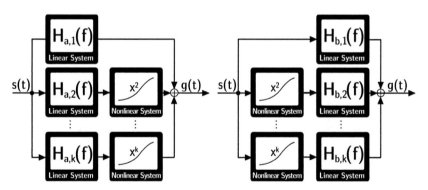

Figure 3.9.: Polynomial Wiener (left) and Hammerstein (right) models [based on (NOVAK, 2009)]

The polynomial Hammerstein model is used by NOVAK in conjunction with measurement results obtained by the exponential sweep method. It is sometimes claimed that only harmonic distortion behavior might be modeled by this approach. As an example two sweeps with the same parameters (length: 12 s, stop margin: 3 s and frequency range 50 Hz to 10 kHz) but only shifted in time by 3 s are exemplarily chosen. These sweeps are then passed through two nonlinear systems with the same characteristics ($x + x^2$, $0\,\text{dBFS}$) and then superposed. This scenario is describing a linear microphone receiving the sound of two nonlinear loudspeakers that each play one sweep at a time. Hence, only the fundamental and the second harmonic of each sweep is received as depicted in Figure 3.10 (left). The right plot shows the result for a nonlinear microphone receiving the superposed sweeps emitted by two linear loudspeakers. In this case, a multitone is present at the input of the nonlinear system and it produces two more lines corresponding to the inter-modulation distortion. A simple nonlinear model produces no inter-modulation only if the input signal is mono-frequent. The second harmonic of the sweep run up to 20 kHz and hence twice of the original frequency.

This example indirectly points to the limitation of the Hammerstein model. A more complex nonlinear system, e.g., a guitar tube amplifier connected to a loudspeaker, where both are driven in a highly nonlinear range, that is assumed to be correctly modeled by two subsequent nonlinear polynomial models—one for the tube amplifier and one for the loudspeaker—could produce inter-modulation for a mono-frequent input signal. This is explained as the first nonlinear block introduces harmonics and the next nonlinear block produces inter-modulation as its input is not mono-frequent anymore. As a consequence, such nonlinear systems cannot be modeled by only one Hammerstein or Wiener model as these models alone cannot introduce inter-modulation to a mono-frequent input signal.

The examples from Figure 3.10 point to another problem concerning Wiener-Hammerstein models and sweeps. Sweeps are considered mono-frequent for small time intervals only. By assuming the first block with memory $(H_a(f))$ to contain an impulse response with two Diracs—one at 0 s and one at 3 s—the same behavior as for the two loudspeaker examples was observed. The signal at the input of the zero-memory nonlinear block cannot be considered mono-frequent and hence the nonlinear block produces inter-modulation although the Wiener-Hammerstein model is fed by a single sweep only (KEMP and PRIMACK, 2011). These inter-modulation artifacts will be analyzed in Section 3.3.4.

3.3.3. Harmonic Impulse Responses using Exponential Sweeps

The output of a nonlinear system to a monochromatic input signal, e.g., a sine with a frequency f_1, is a superposition of pure tones with multiples of this fundamental frequency $f_n = n \cdot f_1$ and $n \in \mathbb{N}$ (FARINA, 2000; MÜLLER and MASSARANI, 2001). For exponential sweeps, which can be considered mono-frequent for short

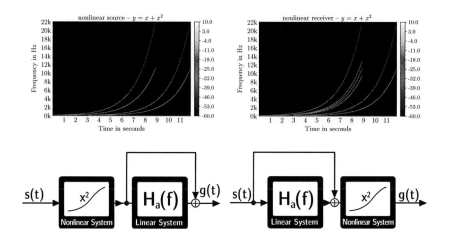

Figure 3.10.: Simulation result (top row) of a scenario with two shifted sweeps passed through two nonlinear loudspeakers and received by a linear microphone (left) and with linear loudspeakers and a nonlinear microphone (right) at 0 dBFS presented as a spectogram with the color indicating the magnitude in dB and the corresponding block diagrams (bottom row), respectively.

time intervals, the output of a nonlinear system is a superposition of harmonics of this sweep.

The dependency of the instantaneous frequency of the harmonics remains exponentially over time. But this slew rate is now a multiple of the original sweep. MÜLLER AND MASSARANI (MÜLLER and MASSARANI, 2001) and FARINA (FARINA, 2000) showed that this results in "time shifted" versions of the original sweep that appear as impulse responses of these harmonics after deconvolution. Hence, nonlinear behavior of the system is observed as anti-causal impulse responses $h_{\mathrm{harm},k}(t)$ for different harmonic orders k separately since frequencies appear in the output before they appear in the input signal (non-causality). This time shift was originally deduced in (FARINA, 2000) by solving:

$$f_{\mathrm{inst}}(t + \Delta t_k) = k \cdot f_{\mathrm{inst}}. \qquad (3.16)$$

By furthermore inserting Eq. 3.11 it holds

$$f_1 \mathrm{e}^{\frac{t + \Delta t_k}{L_{\mathrm{sw}}}} = k \cdot f_1 \mathrm{e}^{\frac{t}{L_{\mathrm{sw}}}}. \qquad (3.17)$$

Dividing by f_1, applying the natural logarithm and resolving to Δt the final results is

$$\Delta t_k = \ln(k) \cdot L_{sw} = \frac{\ln(k)}{r_{sw} \ln(2)} = \frac{\log_2(k)}{r_{sw}}. \qquad (3.18)$$

NOVAK showed that the harmonic impulse responses are generally not in phase with the fundamental impulse response (NOVAK, 2009; NOVAK et al., 2010). This phase shift was neglected before since only a simple time shift was investigated before and it became first important with the characterization of nonlinear system. By further considering the phase ϕ_{sw} the phase offset of the harmonics in relation to the fundamental impulse response can be found as explained later in Section 3.3.6.

The logarithmic modulus of the impulse response of a weakly nonlinear system measured with an exponential sweep is shown schematically in Figure 3.11. Mostly one has interest in the impulse response located to the right in the example in Figure 3.11. From a signal processing point of view, one has of course interest in the *linear impulse response* of the DUT. However, due to measurement method itself the result is in general not directly this linear impulse response of DUT if only even one component of the measurement chain or the DUT itself is driven in a nonlinear range. In many publications, e.g., in (FARINA, 2007; MAJDAK, BALAZS, and LABACK, 2007; MÜLLER and MASSARANI, 2001) the impulse response to the right in the graphic is called linear impulse response which is not correct as can be also seen in the following. The term *fundamental impulse response* is used throughout this thesis in analogy to the terminology used for pure tones.

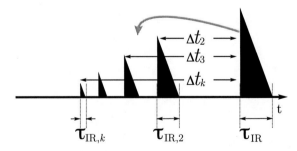

Figure 3.11.: Impulse response of a weakly nonlinear system obtained by exponential sweep measurement where the harmonic impulse respon cede the fundamental impulse response.

Let τ_{IR} be the length of this fundamental impulse response, i.e., the time the system needs to decay into the noise floor. The deconvolved result contains the fundamental and harmonics impulse responses $h_{\mathrm{harm},k}$ with their length $\tau_{IR,k}$. It holds $\tau_{IR,1} = \tau_{IR}$, i.e., the first harmonic is the fundamental itself. The length of the harmonic impulse responses have to be considered as well to avoid interference of the harmonics with the fundamental impulse response. For high sweep rates Δt_2 becomes so small that the harmonic overlaps with the fundamental impulse response. This leads to a constraint for the sweep rate by using Eq. 3.18 and $k = 2$ as follows

$$r_{\mathrm{sw}} \leq \frac{1}{\tau_{IR,2}} \qquad (3.19)$$

where $\tau_{IR,2}$ can be reasonably considered smaller than τ_{IR} for weakly nonlinear systems. Along with the frequency limit of the sweep this constraint requires a minimum length of the sweep τ_{sw} that has to be considered. As the position of the harmonics strongly depends on the sweep rate it is necessary to generate sweeps with exact sweep rates.

For quasi parallel measurement with several weakly nonlinear sources the constraint has to be further narrowed, so that it can be ensured that the harmonics of each source do not overlap with the fundamentals of any source. Originally a measurement method called Multiple Exponential Sweep Method (MESM) was developed that uses replicas of the same sweep delayed by certain times to fulfill this constraint (MAJDAK, BALAZS, and LABACK, 2007). However, this method does not generally provide the shortest delays possible between the sweep as only a subspace of possible solutions for the constraint are found. An advanced method has been proposed that yields potentially shorter delays between the sweeps and hence reduces the overall measurement duration (DIETRICH, MASIERO, and VORLÄNDER, 2013). The sweep constraint to avoid overlapping of harmonics with the part of the fundamental carrying the information of the DUT reads as

$$\tau_{\mathrm{DUT}+} \leq -\frac{\log_2(k)}{r_{\mathrm{sw}}} \bmod \tau_{\mathrm{w}} \leq \tau_{\mathrm{w}} - \tau_{IR,k}. \qquad (3.20)$$

with τ_{w} the time delay between two subsequent sweeps, the sweep rate r_s and the length of the harmonics $\tau_{IR,k}$ and $\tau_{IR,1}$ as the length of the fundamental impulse response. Valid combinations for τ_{w} and r_s for $\tau_{IR,k} = \tau_{IR,1}$ are depicted in Figure 3.12. The shortest measurements durations for a given sweep rate are obtained by using the minimum delay between the sweeps indicated as a bold line in the plot.

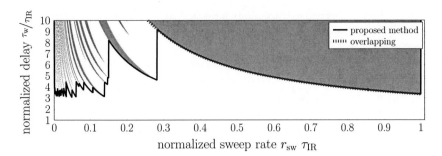

Figure 3.12.: Valid combinations to fulfill the constraints for the proposed method for maximum harmonic order $k_{max} = 4$ and equal lengths of the harmonics and the fundamental in a normalized space (white: interference of harmonics with fundamentals, gray: no interference, black: minimum delay between sweeps) (DIETRICH, MASIERO, and VORLÄNDER, 2013).

For measurements with the entire audible frequency range from 20 Hz to 20 kHz approx. 10 octaves are required. For typical sweep lengths between 0.2 s to 20 s sweep rates between 50 oct./s and 0.5 oct./s are expected.

For some applications it is advantageous to modify the spectral content of the excitation signal for two reasons: Firstly, to adapt to the actual spectrum of the SNR and secondly, to control the maximum amplitude to minimize distortion (HUSZTY and SAKAMOTO, 2010; MÜLLER and MASSARANI, 2001; WEINZIERL, GIESE, and LINDAU, 2009). MÜLLER AND MASSARANI proposed a method to obtain sweeps with arbitrary spectra but constant time envelope. This is realized by manipulation the group delay and hence by using a sweep rate that is frequency dependent. As the resulting instantaneous frequency is not necessarily an exponential function over time, harmonic distortion will no longer appear as sharp impulse responses after deconvolution. Furthermore, distortion artifacts might overlap with the fundamental for some frequencies. Hence, Eq. 3.19 has to be fulfilled for all frequencies, which is not stated in the original publications. This might lead to problems in adapting to the SNR spectra and hence, the adaptation has to be solved in an iterative manner. Shaping the time envelope of a constant envelope sweep can be realized by simply multiplying a zero-phase spectrum. This does not yield in further modification of the position of the harmonic distortion products.

3.3.4. Inter-modulation Artifacts

As sweeps are just assumed to be mono-frequent for a short time interval, an example with a simple nonlinear Wiener model is chosen similar to Figure 3.10 to point out the principal artifacts. Two polynomials are investigated $y_{\text{even}} = x + x^2$ and $y_{\text{odd}} = x + x^3$ to illustrate especially the fundamental difference of these artifacts caused by even and odd orders.

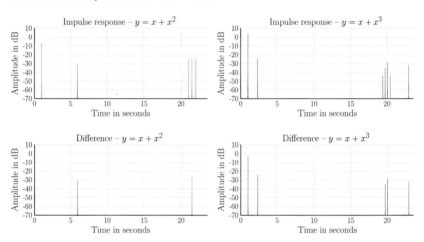

Figure 3.13.: Impulse response (top row) of a nonlinear Wiener system with a double Dirac response ($\Delta t_\text{D} = 1\,$s) placed before the even (top) and odd (right) order polynomial resulting in inter-modulation impulse responses and deviation from two superposed simple nonlinear systems (bottom row) without inter-modulation at $0\,$dBFS.

The impulse response of the linear filter before the nonlinear block is set to two Diracs—one at $0\,$s and one at $\Delta t_\text{D} = 1\,$s. This is obviously an extreme case but clearly shows the separate impulses after deconvolution. The sweep has a length of $23.7\,$s and $r_\text{sw} = 0.365\,$oct./s. The obtained impulse responses for both systems are shown in Figure 3.13. As can be seen, harmonics and additional impulses due to inter-modulation appear in the measured impulse response. These inter-modulation impulse responses are not due to sweeps with integer multiples of the original sweeps as observed before. Two sweeps are observed at the entrance of the nonlinear block—one with the instantaneous frequency f_inst and one with a multiple $k_\text{inter} \cdot f_\text{inst}$ and

$$k_\text{inter} = e^{\frac{\Delta t_\text{D}}{\log_2(e)} r_\text{sw}}. \tag{3.21}$$

For the even polynomial the relative frequency shift Δf_{rel} between the fundamental and the two inter-modulation impulse responses can be calculated by using trigonometric equivalency as

$$\Delta f_{\text{rel,2}} = k_{\text{inter}} \pm 1. \tag{3.22}$$

With Eq. 3.18 follows time difference between the fundamental and the inter-modulation impulse responses based on the relative frequency shift as

$$\Delta t_{\text{inter}} = \frac{\log_2 (\Delta f_{\text{rel}})}{r_{\text{sw}}}. \tag{3.23}$$

The times of arrival require to account for the length of the sweep and Δt_{D} and have to be between $0\,\text{s}$ and the length of the sweep and read as:

$$t_{\text{inter}} = \text{mod}\left(\tau_{\text{sw}} - \Delta t_{\text{inter}} + \Delta t, \tau_{\text{sw}}\right). \tag{3.24}$$

For the given values the impulses are calculated to appear at $5.9\,\text{s}$ and $21.5\,\text{s}$. This matches exactly to the plotted impulse response for the even polynomial. For the odd order the sorted arrival times read as $0\,\text{s}$, $1\,\text{s}$, $2.34\,\text{s}$, $19.74\,\text{s}$, $20.07\,\text{s}$ and $23.98\,\text{s}$. The first two directly interfere with the fundamental. For higher orders, the mixed-terms can be generally calculated using the binomial theorem and subtracting the fundamental and harmonic component (BRONSHTEIN et al., 2005). For a specific order k the following mixed inter-modulation terms appear:

$$(a + b)^k - a^k - b^k = \sum_{i=1}^{k-1} \binom{k}{i} a^{k-i} b^i \tag{3.25}$$

where a and b are pure tones first, that can be later transferred to impulse responses with corresponding normalized frequencies 1 and k_{inter}. For the arrival times, the amplitude of the terms can be neglected. The powers of the pure tones a^{k-i} and b^i can be expressed as a superposition of pure tones with integer multiples η_m—either $\eta \in 1, 3, \ldots, k$ for odd k or $\eta \in 0, 2, \ldots, k$ for even k— of the original frequencies as explained in the following Section 3.3.5. The subscripts m and n are used as counting indices only. The multiplication of these two series can be expressed as a sum of pure tones with normalized frequencies $\eta_m \pm \eta_n \cdot k_{\text{inter}}$.

Furthermore, the difference of the impulse response of the simple Wiener that obviously shows inter-modulation system is compared to a superposition of two nonlinear equal systems with the same polynomial which do not generate inter-modulation. This is similar to the example also given in Figure 3.10. These differences are plotted in the bottom row of Figure 3.13. As can be seen, the odd order polynomial produces not two but four inter-modulation impulse responses. But even more interestingly, the two fundamentals show deviations. These are not the deviation in the fundamental caused by odd order nonlinearities as observed before, as this comparison is against two nonlinear systems that already include the deviation in the fundamental.

It can be concluded that inter-modulation might become problematic if the elements before the nonlinear element in the measurement chain show a combined decay that is long in relation to the sweep parameters, leading to values strongly deviating from $k_{\text{inter}} = 1$. Hence, sweeps with slow sweep rates are less critical and approximate the pure tone case that does not produce inter-modulation. However, inter-modulation in systems with odd orders is problematic as it affects the fundamental itself and also the range around the fundamental. This might lead to pre-ringing and post-ringing for all symmetric distortion (KEMP and PRIMACK, 2011). All systems that incorporate a Hammerstein model are potentially affected. In a room acoustic measurement chain, this is typically fulfilled as long as the loudspeaker is the nonlinear part and not the microphone where a linear filter with long decay would precede the nonlinear block. The loudspeaker itself can be reasonably modeled to consist of a short impulse response—a short Δt_{D} in the previous example—before the nonlinear polynomial. Therefore, inter-modulation artifacts with moderate levels and moderate sweep rates fulfilling the requirements above are not to be expected.

3.3.5. Relationship between Harmonics and Polynomial Coefficients

In general, the harmonics and particularly its amplitudes depend on the level of the input signal whereas the polynomial coefficients describe the system behavior for all levels. In this first step, unity amplitude is assumed. Based on the findings in (NOVAK, 2009) the appearance of harmonics for systems described by its polynomial coefficients can be formulated. In the following this formulation is adapted and enhanced to correctly account for the level of the input sines and the phase of the output. Basic substitutions for powers of cosines with an amplitude of 1 are known, e.g., $\cos^2(x) = 1/2(1 + \cos(2x))$ or $\cos^3(x) = 1/4(3\cos(x) + \cos(3x))$. Hence, powers of cosines with a specific frequency might be written as a superposition of cosines of multiples of this frequency and a constant. The formulation for arbitrary higher orders can only be given for even and orders separately. Hence, a different approach is used in the following.

If a cosine is passed through a CHEBYSHEV polynomial $T_{p,k}$ of order k the output is once again a cosine function but with k times the input frequency:

$$T_{p,k}\left(\cos\left(\omega t\right)\right) = \cos\left(k\omega t\right).\tag{3.26}$$

This relation for cosines of unity amplitude can be compactly written in matrix notation

$$h_c = \mathbf{CM}^{-1} \cdot p_c\tag{3.27}$$

where h_c is the vector containing the harmonic coefficients starting at order 0 up to the maximum harmonic order, p_c is the vector of the polynomial coefficients and $\mathbf{CM}$ a matrix containing the CHEBYSHEV polynomial coefficients C_k. The harmonic coefficients for the substitution of a cosine of order k can be calculated by inserting a vector p_c with a 1 at the position for the kth order and zeros elsewhere. The coefficients of the CHEBYSHEV polynomials are defined by recursion (BRONSHTEIN et al., 2005)

$$\begin{aligned}
C_0(x) &= 1 \\
C_1(x) &= x \\
C_k(x) &= 2 \cdot C_{k-1} \cdot (x) \cdot x - C_{k-2} \cdot x
\end{aligned}\tag{3.28}$$

and might be summarized in a column vector $T_k = (C_0, C_1, \ldots, C_k)^{\mathrm{T}}$ with increasing order of appearance for the coefficients for the order k.

The matrix $\mathbf{CM}$ up to an order k can be generated by placing the coefficient vectors T_i with increasing order i of the CHEBYSHEV polynomials:

$$\mathbf{CM}_k = (T_0, T_1, \ldots T_i, \ldots T_k).\tag{3.29}$$

With this approach the polynomial coefficients of a frequency independent non-linear systems might be determined. As shown in the following, a prediction due to a change in level is not yet included and the phase relation of the harmonics obtained by sweep measurements instead of cosines has not been captured, yet.

3.3.6. Generalization of the Relation

As the relation is valid for cosines of unity amplitude only a generalization is deduced in the following for cosines of arbitrary phase and finally of cosines with arbitrary starting phase and exponential sweep measurements.

Amplitude and Scaling Correction

The gain G of the input signal has to be considered for each polynomial order separately[4]. E.g., a cosine of amplitude $1/2$ passed through a system with $y = x^2$ will produce $1/4$ of the result obtained with a cosine of unity amplitude. For exponential sweep measurements the gain G is always compensated within the deconvolution process. By combining both findings the relation generalizes to

$$h_{\mathrm{c}} = \mathbf{CM}^{-1} \cdot \mathbf{GM} \cdot p_{\mathrm{c}} \qquad (3.30)$$

with the diagonal *gain matrix* $\mathbf{GM}$

$$\mathbf{GM} = \begin{pmatrix} G^{-1} & 0 & 0 & 0 \\ 0 & G^0 & 0 & 0 \\ 0 & 0 & \ddots & 0 \\ 0 & 0 & 0 & G^{\prime k-1} \end{pmatrix} \qquad (3.31)$$

including also the DC component in the first row and the first column.

The simulated energies—by using a polynomial model as introduced before—for the harmonics for simple polynomials are depicted in Figure 3.14 (dashed line) according to the relation found in Eq. 3.30. The first harmonic or fundamental does not depend on the level of the input signal if the system is linear $(y = x)$ as the gain of the input signal is already accounted for by the deconvolution. As can be seen, the inclination of the straight line for one polynomial order is equal for all harmonics generated by this order and is $(k - 1) \cdot 1\,\mathrm{dB}$ per input level in dB. Furthermore, even polynomial orders generate even harmonic orders only and the same holds for odd orders. The difference in the inclination of the level of the harmonics, e.g., caused by $y = x^6$, is important to mention at this point. Even though the system only contains one polynomial order the even harmonics of lower order have more energy than the harmonic of that particular order 6.

The results (continuous lines in Figure 3.14) obtained by passing an exponential sweep through an emulation of such a nonlinear systems with the same parameters and then separating the harmonic impulse responses closely matches the theoretical results found before. Deviations are due to a limited frequency range of the exponential sweeps causing the fundamental as well as the harmonic impulse responses to smear slightly into each other. These deviations are more likely to be found for lower input levels and have low values only.

---

[4]$G = 1$ corresponds to $0\,\mathrm{dBFS}$

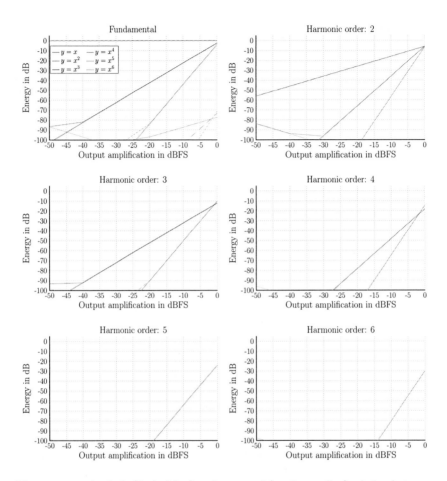

Figure 3.14.: Analytic (dashed line) and measured (continuous line) relation between input level in dBFS and level of specific harmonics for systems with simple polynomials.

As a conclusion of the observed relationship higher order harmonics can be significantly reduced in amplitude if the level of the excitation signal is slightly reduced.

Sweep Rate and Phase Correction

For cosines $\cos(2\pi f t + \psi_{\text{offset}})$ with starting phase ψ_{offset} the matrix relation can be written as

$$h_{\text{c}} = \mathbf{PM} \cdot \mathbf{CM}^{-1} \cdot \mathbf{GM} \cdot p_{\text{c}} \qquad (3.32)$$

with the *phase matrix* $\mathbf{PM}$. The ith harmonic has a phase offset by $i \cdot \psi_{\text{offset}}$ and hence, the diagonal phase matrix $\mathbf{PM}$ comprises of exponents of a phase offset ψ_{offset} similar to the gain matrix before as

$$\mathbf{PM} = \begin{pmatrix} e^{-1 \cdot j \psi_{\text{offset}}} & 0 & 0 & 0 \\ 0 & e^{0 \cdot j \psi_{\text{offset}}} & 0 & 0 \\ 0 & 0 & \ddots & 0 \\ 0 & 0 & 0 & e^{(k-1) \cdot j \psi_{\text{offset}}} \end{pmatrix} \qquad (3.33)$$

including again the DC component in the first line and the first column.

As known for exponential sweep measurements the phase of the fundamental is already correct due to deconvolution. But the phase offset is sligthly more complicated than for pure cosines as this offset for exp. sweep measurements itself contains two parts. The first factor is caused by the variable starting phase of the exp. sine sweep as expressed in Eq. 3.13. The second factor of $\pi/2$ converts from sines to cosines as used in the Chebyshev relation before.

$$\psi_{\text{offset}} = \phi_0 - 2\pi f_1 L_{\text{sw}} - \frac{\pi}{2}. \qquad (3.34)$$

Although the necessary equations to obtain such a phase offset are already deduced in (NOVAK, 2009) they are only used to generate a specific class of sweeps that does not generate phase offsets by manipulating the sweep rate and forcing the constant phase offset in Eq. 3.13 to zero. This results in a strong limitation of possible sweep lengths or the frequency range.

A simulation of a nonlinear system with the polynomial $y = x + x^2 + x^3 + x^4$ with an exp. sweep ($f_1 = 5\,\text{Hz}$, $r_{\text{sw}} = 1.0483\,\text{oct.}/\text{s}$, $\phi_0 = 1$) with a driving amplitude of $G = 1/2$ is analyzed exemplarily. The predicted amplitudes by using Eq. 3.30 are 1.1875, 0.3125, 0.0625, 0.0156 for the fundamental, the second, third and fourth harmonic, respectively The phase offset ψ_{offset} can be calculated as -30.54 in radians. The resulting phase shifts of the harmonics (2,3 and 4) can be given as $50.2°$, $100.4°$ and $150.6°$, respectively by using Eq. 3.32. The results using the emulation of the measurement chain are shown in Figure 3.15 for the amplitude,

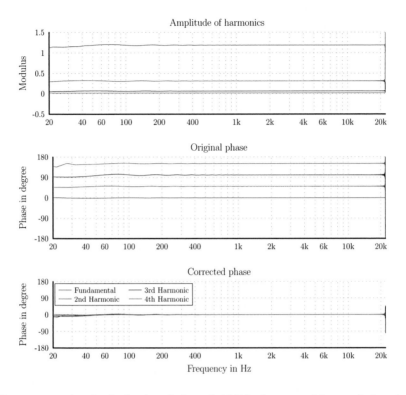

Figure 3.15.: Amplitude (top) and phase (middle) of separated harmonic impulse response from an exp. sweep measurement and corrected phase (bottom) by subtraction calculated phase offsets due to sweep rate and starting phase of the exp. sweep.

the measured phase and corrected phase, by subtracting the predicted phase offsets for the harmonics.

Although these results are very promising for the characterization of nonlinear systems as the problem with the phase offset seems to be solved still some problems remain. As can be seen in the plots, a slight ripple in both amplitude and phase can be observed which is caused by the finite length and band limitation of the sweep. The predicted phase offset is very sensitive to small errors in the sweep rate. This problems occurs mainly when $\psi_{\text{offset}} \gg 2\pi$. A small relative error will result in a large relative error in the predicted phase corrections due to the modulus operation with 2π. This is increasing for increasing harmonic orders. It is worth mentioning, that the same problem also occurs for the class of sweeps used by NOVAK. Sweeps can be generated with very accurate sweep rates by directly using Eq. 3.13. However, this problem should be checked carefully for the characterization of nonlinear systems.

3.4. Implementation and Emulation of the Measurement Chain

The nonlinear model described above is used inside the measurement chain as depicted in Figure 3.16 to emulate measurements. This measurement chain is used in this chapter for detailed investigations of the uncertainties introduced by a nonlinear element.

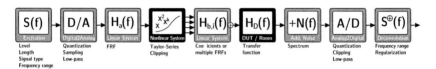

Figure 3.16.: Block diagram of the measurement chain as implemented in the ITA-Toolbox using MATLAB for the emulation of transfer function measurements including noise, quantization, nonlinear and linear transfer characteristics.

General Structure

The structure of the measurement chain is based on typical measurement chains as also shown in Figure 3.4 in acoustics and hence the nonlinear block (typically loudspeaker or shaker) is placed before the linear transfer function of, e.g., a room. The parameters controllable by the user for each block are listed below each block. The effects theoretically introduced by each block were described in Section 3.2.

Time-discrete Problems

In case of time discrete signals the exponentiation cannot be directly applied as written for the continuous case, e.g., in textbooks. The exponentiation is a multiplication of the time signal with itself and can better be understood in frequency domain as a convolution. Analog to the examples already presented for time-aliasing in Section 2.2.3 or for the cyclic deconvolution aliasing in frequency domain occurs. Hence, to avoid these aliasing artifacts two possibilities exist. Firstly, the signal might be low-pass filtered with appropriate setting of the upper frequency limit to the NYQUIST frequency divided by the exponent or secondly, by using oversampling with a factor equal to the exponent. The latter method has higher computational complexity but minimizes further artifacts, e.g., due to filtering or band-limitation.

The oversampling variant can simply be implemented by zero-padding in frequency domain and exponentiation of the time discrete signals in time domain, followed by down-sampling in the end to obtain the same sampling rate as for the input signal. This approximates linear convolution in frequency domain, analog as for the linear deconvolution earlier. As an example, the spectrogram of an exponential sweep passed through a modeled system with $y = x^5$ is shown in Figure 3.17 for direct exponentiation and by using appropriate oversampling.

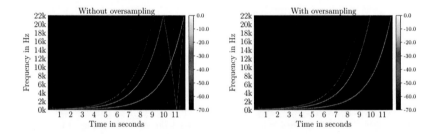

Figure 3.17.: Comparison of time discrete nonlinear model with a polynomial $y = x^5$ applied in time domain with and without using an oversampling technique to avoid aliasing.

In this context the MLS signal should be considered. Without applying such anti-aliasing filtering the response of a simple nonlinear block with an odd order polynomial of the type $x = x^k$ to an MLS is the same MLS. The response to even order polynomials of the same type would result in a constant in time domain. Hence, the specific spikes could not be observed and hence such nonlinear modeling would not comply with realistic nonlinear systems.

Polynomial Wiener or Hammerstein Model

To model the nonlinear behavior in more detail the generalization towards the polynomial Wiener or Hammerstein model can be applied. These polynomial models use a simple split of the signal for specific polynomial orders preceded (Wiener) or followed (Hammerstein) by the conventional convolution from LTI theory and final superposition of these signals.

3.5. Dealing with Nonlinear Systems

Besides the criterion to avoid overlapping of the harmonics with the fundamental impulse response more advanced signal processing methods can be applied to cope with nonlinear systems.

3.5.1. Separation of Harmonics

In theory, the separation of the fundamental and each harmonic impulse response out of the measured impulse response can be ideally realized using an infinitely short time window around the impulse response. The time of occurrence of each harmonic can be calculated easily from the sweep parameters according to Eq. 3.18. However, real systems show a decay and the fundamental and harmonic impulse responses spread in time. Hence, the window has to be adapted accordingly. The time between the harmonic impulse responses decreases with order, leading to a trade-off between capturing the information of the harmonics with long decay and hence minimizing cross-talk from the neighboring harmonic impulse responses and the measurement duration.

For the correct identification of a nonlinear system or the calculation of its total harmonic distortion at a certain level each harmonic up to an order k has to be determined with sufficient SNR and sufficient length of each harmonic. But in addition to the constraint for the sweep rate in Eq. 3.19 more constraints have to considered for each harmonic order to avoid overlapping of harmonics. It is assumed that the harmonic impulse responses have the same decay and hence the same length to be considered is the same as of the impulse response obtained in a straight linear range. It is sufficient to claim a minimum time interval between the harmonic of highest order k and the preceding harmonic as follows:

$$\Delta t_k - \Delta t_{k-1} = \frac{\log_2(k) - \log_2(k-1)}{r_{\mathrm{sw}}} \geq t_{\mathrm{RIR}}. \tag{3.35}$$

Hence, the sweep rate has to be below the following limit

$$r_{\mathrm{sw}} \leq \frac{\log_2(k) - \log_2(k-1)}{t_{\mathrm{RIR}}} \tag{3.36}$$

and hence the measurement duration increases disproportionate with increasing harmonic order to be considered.

The harmonic impulse responses of a nonlinear system driven with moderate amplitudes are far below the level of the fundamental impulse responses as already shown in Figure 3.14. As the noise floor in measured impulse responses is usually constant over time of the impulse response the SNR of the harmonics is far below the SNR of the fundamental impulse response. Hence, the SNR criterion has to be fulfilled for each of all harmonic orders of interest. A practical example with $k_{\mathrm{max}} = 5$ and a relative level of the fifth harmonic of $-50\,\mathrm{dB}$ compared to the fundamental impulse response, requires the SNR to be improved by these $50\,\mathrm{dB}$ to obtain the same quality as originally measured for the fundamental

In context of uncertainty analysis in this thesis the identification of nonlinear systems is not further considered. However, it can be stated that due to the complexity and the problems observed by determining the parameters for a Wiener-Hammerstein model, the uncertainty in the separated harmonic impulse responses, e.g., due to limited SNR, might lead to large errors in the model. A simple correction of the measured fundamental impulse response using the nonlinear information stored in the harmonics to obtain the *linear impulse response* of the system does not seem to be practical at this point.

3.5.2. Suppressing certain Harmonics

In some practical applications harmonic impulse responses might interfere with the fundamental impulse responses. Hence, a method to suppress harmonics is required if the windowing technique explained in the previous section is not applicable. Using Eq. 3.15 and the fact that even harmonics are associated with even exponents in this equation, a simple but effective technique can be developed. The fundamental impulse response obtained with a phase shifted sweep of same amplitude remains the same, but some harmonics are shifted in phase. In particular, the harmonics of even order are shifted by $180°$ when the sweep is shifted by $180°$. Let the result $g_1(t)$ at the output of the nonlinear system to the sweep $s(t)$ be

$$g_1(t) = a_1 s(t) + a_2 s^2(t) + a_3 s^3(t) + a_4 s^4(t) + \dots \qquad (3.37)$$

and the impulse response $h(t)$ after deconvolution with $s(t)$ be

$$h_1(t) = x_1(t) + x_2(t) + x_3(t) + x_4(t) \dots \qquad (3.38)$$

where $x_i(t)$ are the harmonic impulse responses of ith order. The system output $g_2(t)$ for an inverted excitation signal $-s(t)$ is

$$g_2(t) = -a_1 s(t) + a_2 s^2(t) - a_3 s^3(t) + a_4 s^4(t) + \dots \qquad (3.39)$$

but due to the deconvolution the phase is shifted by $180°$

$$h_2(t) = x_1(t) - x_2(t) + x_3(t) - x_4(t) \dots \quad . \qquad (3.40)$$

Hence, a superposition of the two different measurements suppresses all even harmonics, regardless of their temporal position in the impulse response.

A practical example measured with a custom-made mid-frequency dodecahedron loudspeaker in the hemi-anechoic chamber is given in Figure 3.18. The excitation level has been chosen fairly high to obtain clearly visible harmonics and hence, the SNR is very high. The impulse response obtained by a classical measurement using one exponential sweep shows harmonic impulse response at least until the order of 13. Due to proposed combination or averaging with two exponential sweeps of inverted phase even order harmonics are removed and as a consequence of the averaging the SNR is further increased by $3\,\mathrm{dB}$. By subtracting both measurement results and dividing the resulting impulse response by a factor of 2 (the same is applied for the averaging before) the odd orders can be suppressed instead with the same SNR as for the averaging. Minor deviations in the fundamental part can be observed that are assumed to be caused by time variance in the loudspeaker system as it was driven with very high amplitude and the time interval between both measurements was to short for, e.g., cooling down of the voice coil.

It is theoretically possible to suppress harmonics of more orders than only the even or odd orders at the same time. Averaging four impulse response measurements with phase shift of $90°$ for the exponential sweeps suppresses all orders except for the orders $k = 4 \cdot N + 1$. As the number of harmonics usually observed in

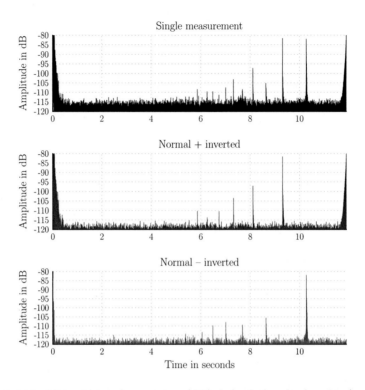

Figure 3.18.: Measured impulse response of ITA dodecahedron loudspeaker (zoomed y-axis, maximum peak is around 0 dB) in the hemi-anechoic chamber for single, normal measurement (top) and averaging of two measurements with inverted phase (middle) suppressing even harmonics and averaging by using subtraction (bottom) to suppress all odd orders.

practical acoustical measurements is limited and typically below $k = 8$ eight measurements with a phase offset of $360°/8 = 45°$ is sufficient to suppress all harmonics in the measurements. However, such a phase shift results in sweeps without a smooth beginning and end and not considered very practically. A similar approach was found to be published as a patent for seismic applications in (MOERIG et al., 2004).

It can be concluded that in case averaging is already applied in a measurement with an even number of averages this simple technique to suppress harmonics of even orders should always be considered as no further errors are introduced.

3.5.3. Measuring at the Quantization Limit

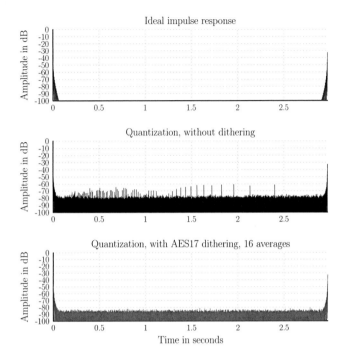

Figure 3.19.: Simulated quantization artifacts in the ideal impulse response obtained with exponential sweep at $-100\,\mathrm{dBFS}$ without quantization (top), with quantization (24 bit) without (center) and with $AES17$ dithering (bottom).

Quantization effects are commonly modeled as additive white Gaussian noise. This assumption is valid if the signal level is far above the quantization level, e.g., more than $40\,\mathrm{dB}$. If one zooms more into detail, i.e., the signal amplitude is lowered or the quantization steps are increased, the nonlinear transfer characteristics of the ADC becomes evident. But more interestingly, quantization is also applied to fine floating point representations (e.g. 64 bit) of time discrete signals before arriving at the DAC of the sound card as fixed point representations are used in audio hardware (e.g. 24 bit).

Due to the quantization the SNR is limited regardless on the level of the excitation signal. Even applying averaging cannot further lower the noise floor in the impulse responses as the quantization artifacts are always correlated with the excitation signal if the same excitation signal is used for all measurements. As described in Section 2.2.5 dithering might be applied to break the strong correlation of the quantization artifacts with the excitation signal. Hence, applying dithering to the excitation signal with different dithering noise for each measurement can increase the SNR.

A simulation with the emulated measurement chain illustrates the behavior with and without dithering in Figure 3.19. An exponential sweep with 6 s and a frequency range of 100 Hz to 22.5 kHz with a stop margin of 0.1 s is used. The quantization is applied with 24 bits and the excitation signal is 100 dB below the maximum amplitude of the quantizer.

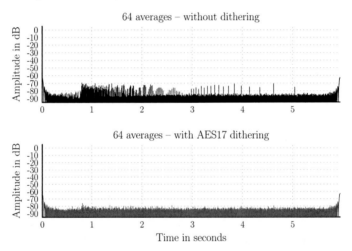

Figure 3.20.: Measured quantization (24 bit) artifacts in the impulse response of an *RME Multiface* with output connected directly via optical TOSLINK to the input without (top) and with *AES17* dithering (bottom).

A practical example measured with an *RME Multiface* using an optical Toslink connection between ADAT input and output is given in Figure 3.20 for 64 averages. The resulting impulse response without dithering is very similar to the simulated impulse response in Figure 3.19 before. Due to different sweep lengths (approx. 3 s and 6 s) used, the measurement example reasonably shows 3 dB more SNR. The simulation is considered to be accurate enough to predict the behavior observed in the real life measurement. As mentioned before, averaging without dithering is not expected to increase the signal to noise ratio. Hence, the impulse response for 64 averages equals the impulse response for a single measurement if the sound card or its software driver does not apply dithering itself as it was the case in this measurement. A comparison with the simulated impulse response shows good agreement as the quantization artifacts (spikes) disappear in the noise floor. Hence, it is possible to measure at the quantization limit or beat the theoretical SNR limit caused by the quantizer. But considering the number of averages it becomes obvious that a higher number of averages is required to improve the result significantly.

Concluding for practical measurements of acoustical systems the quantization effect only becomes prominent at very low excitation levels and is considered mainly irrelevant for, e.g., room acoustic measurements. However, measurements using a high number of averages might potentially benefit from dithering and hence, dithering should be applied in these cases.

3.6. Post-processing of Measurement Data

There exist various post-processing methods for measured impulse responses, varying in their purpose and complexity.

The most important post-processing step is the equalization or compensation of the equipment used. This requires detailed knowledge of especially the FRF of the LTI components of the equipment in the measurement chain. The uncertainties in the FRFs of the components have a direct effect on the uncertainty of the FRF of the DUT. E.g., a frequency independent level calibration error of $-1\,\mathrm{dB}$ results in an error of $+1\,\mathrm{dB}$ in the final result due to the inversion with the FRFs of the elements of the measurement chain. An example from a room acoustical analysis of the lateral fraction (LF) due to a level calibration uncertainty has been investigated and published in (DIETRICH and I. WITEW, 2008). The uncertainties of the sensitivities of two different microphones used for this kind of measurement remain. As the other elements of the measurement chain stay the same in both measurements the influence vanishes.

When the purpose of a measurement is to serve as a graph in a presentation the post-processing has to be obviously chosen to generate a smooth and nice looking curve. Frequency smoothing with either a fixed absolute bandwidth in Hertz or a relative bandwidth in fractions of an octave are applied in that case. This approach is applied in frequency domain by using the weighted average of the frequency data points in a region. It reduces observable noise influences or the small variations in frequency curves but does not necessarily improve the quality of the measured impulse response. In fact, valuable information might be treated as noise as no *a priori* knowledge about the DUT's response is used. This approach is not suggestible for technical analysis of the behavior of a system and therefore not considered further.

The most common post-processing method is to apply a fractional octave band analysis. The uncertainty of the levels of each band scales reciprocal to the filter bandwidth used as fewer frequency bins are available with decreasing bandwidth. But the information is reduced significantly as well. Especially phase information is discarded.

The multiplication of the measured impulse response with a time window is a more powerful approach than the both aforementioned approaches. A priori information on the temporal structure of the impulse response is used to suppress unwanted reflections, echo, reverberation, noise or artifacts as, e.g., harmonic impulse responses as already investigated. By considering, e.g., a reflectogram with perfect Dirac impulse a simple rectangular time window might be applied to single out a certain reflection without introducing any errors. But in case a real world subsystem with high-pass characteristics (e.g. studio sound cards or loudspeakers) is used somewhere in the measurement chain this perfect Dirac impulses become wider. A problem arises as, e.g., the rectangular window suddenly produces discontinuities in the post-processed impulse response of the DUT. Hence, different window functions are applied to minimize this effect. A second problem is the length of the portion to be windowed out as required information might be smeared out of the range of this window. In this case an increase of the length of the window improves the lower frequency limit. This last effect is often applied in a general rule of thumb by pointing out a lower frequency limit just due to the window length only. It is in fact a more complex relationship. For the perfect Diracs the lower frequency limit due to this post-processing step remains 0 Hz although a finite window length is used. As an improved rule of thumb one can say that the variation in frequency domain at low frequencies of the single components to be separated by the time window determines the lower frequency limit along with the window length. No variation as in case with the Diracs works perfectly with all kinds and lengths of time windows as long the window is placed in a manner that, e.g., unwanted reflections are still suppressed. If the temporal structure of the impulse response just comprises of the DUT's response and noise the window is used to suppress the noise only.

The same variation of the frequency curve for high frequencies results in a faster decay as for low frequencies. This effect can be used to define more sophisticated time-windows, that are frequency dependent. While there exist methods involving the time-frequency domain a simple but effective method was developed during measurements for this thesis. The measured impulse response can be divided into sub-bands with the constraint that the superposition of all sub-bands is able to reproduce the original data, e.g., by *Linkwitz-Riley filters*. These sub-bands are then processed with different window parameters, e.g., shorter windows for higher frequencies. A different but very similar approach is to apply time windows different with different parameters to the broad-band impulse response and cross-fade the results in frequency domain. The latter method is not only simple to implement but also has less computational complexity.

3.7. Application—Uncertainties in Room Acoustic Parameters

An example from the field of room acoustics is studied by using a simulation approach. Generally, room acoustic parameters are calculated based on measured or simulated impulse responses as defined in ISO 3382 (ISO 3382, 2009). In the following the influence of nonlinearities introduced by a loudspeaker and hence leading to uncertainties in the impulse responses are examined with a simple nonlinear model as described above. Noise effects as, e.g., published in a summarizing manner in detail for practical measurements in (GUSKI and VORLÄNDER, 2013) and quantization effects are not considered at this point. However, the developed emulated measurement chain is capable of delivering quite realistic results comparable to the published results.

3.7.1. Simulation Setup

The ideal transfer function is obtained by modal superposition as later described in Section 4.3.1 with a mean reverberation time of approx. 1 s and a room geometry of $8 \times 5 \times 3$ meters up to a frequency of 4 kHz. This approach allows for arbitrary decays and noiseless input data[5]. The ideal impulse response is

---

[5]Parts of the following results have already been published in (DIETRICH, GUSKI, and VORLÄNDER, 2013).

shown in Figure 3.21. For the ideal room impulse response the parameters EDT and C_{80} are depicted in Figure 3.22[6].

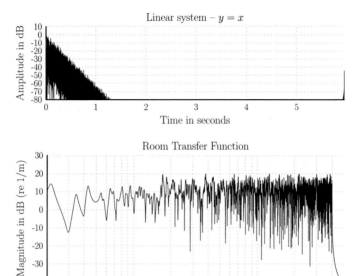

Figure 3.21.: Ideal room impulse response (top) and room transfer function (bottom) obtained by modal superposition with a simple analytic model for rectangular rooms (modes calculated up to 4 kHz) used in the emulation of the measurement chain.

To study the influence of even and odd orders separately the two polynomials $g\left(t\right)_{\mathrm{even}} = s(t) + s^2(t)$ and $g\left(t\right)_{\mathrm{odd}} = s(t) + s^3(t)$ are chosen exemplarily. The *input-output-diagram* for both polynomials is depicted in Figure 3.23. For higher input levels the deviation from linearity can be observed. Higher orders are not used due to a low benefit to the demonstration of the artifacts. For exponential sweep measurement two different artifacts are expected for the even and odd orders nonlinearities. A level of 0 dBFS yields linear and nonlinear terms with identical energy. This is chosen as a worst case scenario. The level of total harmonic distortion is typically far below 10 % and hence even extreme violations of the linear range are captured in the following results. The relation between

---

[6]Details on the derivation of the room acoustic parameters from the impulse responses are provided in Section 4.3.2.

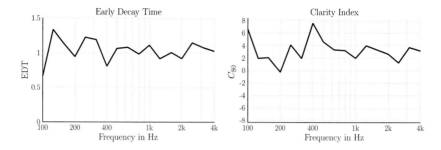

Figure 3.22.: Room acoustic parameters EDT and C_{80} obtained from simulated ideal room impulse response in third-octave bands.

THD and the output amplification for the two exemplary polynomials calculated according to Eq. 3.1 and Eq. 3.30 is depicted in Figure 3.23.

The sweep rate has been chosen in a way that the second harmonic (even order) partly overlaps with the fundamental impulse response and hence violates Eq. 3.19 but the third harmonic (odd order) does not overlap. The length of the sweep is 4 s followed by a silence of 2 s. The sampling rate was 44100 Hz and the frequency range of the sweep was 100 Hz to 16 kHz. Hence, the sweep rate was 1.9 oct./s. The influence due to this overlapping can be studied by varying the amplitude of the sweep noted in dBFS in the plots. For increasing amplitudes the second harmonic will increase in relation to the fundamental impulse response (c.f. Figure 3.14).

As mentioned earlier the fundamental impulse response—besides this overlapping—is influenced by odd polynomial orders only. As long as the level is kept constant between two measurements the fundamental is affected in the same way in both measurements and deviations might cancel out by, e.g., by spectral division (MÜLLER and MASSARANI, 2011). By using the odd order polynomial $g_{odd}(t)$ this effect can be modeled with a variation of the driving amplitude at the input of the nonlinear model as well.

The driving amplitude of the sweep (*output amplification*) is increased step-wise to study the increasing influence of the nonlinearities. The resulting impulse response obtained by the emulated measurement for 0 dBFS and for the even order polynomial is shown in Figure 3.24. As can be seen, the deviation compared to the ideal impulse response due to overlapping in the beginning of the simulated impulse response is approx. 40 dB below the level of the ideal impulse response.

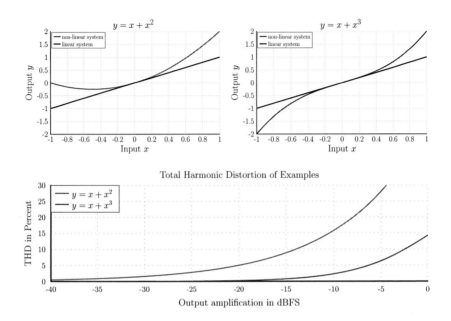

Figure 3.23.: Input-output-diagram for $g(t)_{even}$ and $g(t)_{odd}$ used as two simple non-linear models (top). Total harmonic distortion over output amplification for the two exemplary polynomials (bottom).

Due to this small deviation only small deviations of the room acoustic parameters might be expected if the overlapped signal is considered as noise.

For the odd orders, the impulse response of the third harmonic can be clearly seen in the end of the impulse response in Figure 3.25. But more interestingly, the fundamental impulse response has also changed as can be seen in the difference plot. The level of the deviation is almost as high as the ideal impulse response itself. Hence, errors in the room acoustic parameters might be expected. To investigate the spectral deviations for different output amplifications the simulated impulse responses are time windowed to separate the fundamental from the harmonic impulse responses at around 1.75 s. The spectrum of these fundamentals for the different output amplifications is referenced to the spectrum of the linear impulse response and shown in Figure 3.26. For low output amplifications and hence low distortion the deviations are below ± 0.1 dB but the modal structure can also been observed in the errors for the even polynomial. For the odd order polynomial the part of the error that shows the modal structure is constant for all

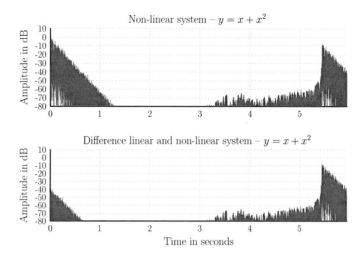

Figure 3.24.: Simulation of impulse response with emulated measurement chain using the nonlinear model $g(t)_{\text{even}}$ (top) at $0\,\text{dBFS}$ with a strong second harmonic and deviation from ideal impulse response (bottom).

amplifications. This is caused by a remaining overlap of the third harmonic with very low level. The constant deviation over frequency is caused by the influence of the third order term into the fundamental as described before.

3.7.2. Derivation of Room Acoustic Parameters

Room acoustic parameters are calculated for each impulse response obtained for the different output amplifications. No noise handling or correction methods methods were applied. The parameters were calculated for each frequency band by applying a third octave band filter according to ISO 3382 prior to the evaluation.

The error in the room acoustic parameters due to a more and more overlapping second harmonic is depicted in Figure 3.27 for the Early Decay Time (EDT) and the clarity index C_{80}. For the EDT the error is fairly small and similar for all third-octave frequency bands except for the frequency band around $100\,\text{Hz}$. This can be explained by a small dip in the room transfer function at this frequency and hence the overlapping signal has a higher impact. Although the EDT is known to be very sensitive to variations due to a short portion of the impulse response used for evaluation, the artifacts due to overlapping are not that dramatic even

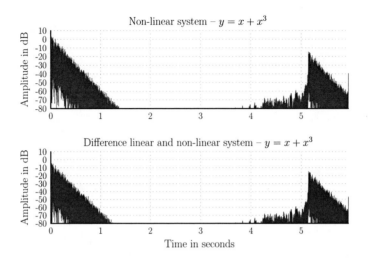

Figure 3.25.: Simulation of impulse response with emulated measurement chain using the nonlinear model $g(t)_{\mathrm{odd}}$ (top) at $0\,\mathrm{dBFS}$ and deviation from ideal impulse response (bottom).

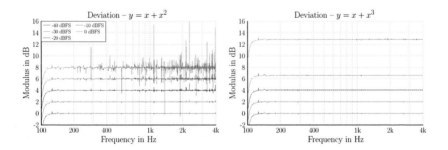

Figure 3.26.: Spectral deviation for even (left) and odd (right) order polynomial from the ideal FRF. Results are shifted by $2\,\mathrm{dB}$.

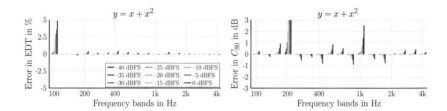

Figure 3.27.: Error in the room acoustic parameters EDT (left) and C_{80} (right) for different levels of the excitation signal and *even* order polynomial.

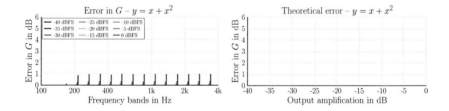

Figure 3.28.: Error in the room acoustic parameter sound strength G derived from the simulated impulse response (left) and theoretical error (right) for different levels of the excitation signal compared to the ideal linear results and *even* order polynomial.

for very high levels. The errors in the clarity index show a completely different behavior. Although the errors increase with increasing output amplification as expected, the magnitude of this error is different for the frequency bands.

Figure 3.28 (left) shows the error in the sound strength including the simulated impulse response. This error is increasing with increasing output amplification and similar for all frequency bands above 200 Hz. The frequency response of the second harmonic in the measurement only contains frequencies above 200 Hz which is exactly twice ($k = 2$) the frequency limit of 100 Hz in the measurement setup. The error due to a simple change in level of the fundamental without using the impulse response in the emulated measurement chain is independent on frequency and shown in Figure 3.28 (right) over the output amplification. For the even order polynomial no error in magnitude of the fundamental impulse response and hence no error in the sound strength is theoretically expected as no overlapping occurs which corresponds with the plotted results.

Due to the overlapping, errors in the room acoustic parameters can be observed. As mentioned earlier the energy of the overlapping harmonic had 40 dB less energy than the fundamental. Hence, the sweep rate should be always chosen to avoid such overlapping by considering Eq. 3.19 since the errors could be in the same order of magnitude as the Just Noticeable Difference (JND) of approx. 5 % for reverberation times (only observed for one frequency band in the example) and 1 dB for the clarity index (ISO 3382, 2009). By claiming THD to be smaller than 10 % the errors are much smaller, but they might still exceed the JND for some case as observed for the clarity index.

The deviation of the room acoustic parameters EDT and C_{80} obtained from the odd order nonlinear model are shown in Figure 3.29 in the same manner as for the even order. The errors are much smaller than for the even order polynomial as virtually no overlapping occurs. The magnitude of the errors in these parameters is far below the JND even for the extreme cases with very high distortion at 0 dBFS. This seems contradictory to the impulse response shown before in Figure 3.25 as the fundamental shows high deviations. However, the deviation is just a frequency independent level shift due to the frequency independent model. The error in sound strength is shown in Figure 3.29 (left). As can be seen, the order of magnitude for moderate distortion is smaller than 1 dB but might superpose with other uncertainties to values above 1 dB. The errors are very similar for all frequency bands and also below 300 Hz ($k = 3$). This can be explained as third harmonic will generally start at frequencies above three times 100 Hz but the artifacts in the fundamental due to the term x^3 starts already at 100 Hz.

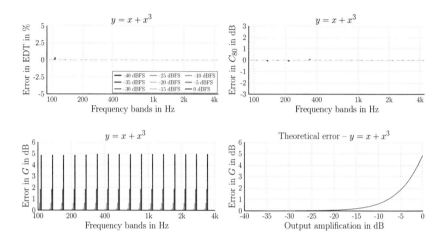

Figure 3.29.: Error in the room acoustic parameters EDT (left) and C_{80} (right) for different levels of the excitation signal and *odd* order polynomial (top). Simulated and theoretical error in sound strength G.

3.7.3. Comparison with Measurement Results

A measurement was carried out in the large auditorium *Aula I* at RWTH Aachen University at several positions. The mid-frequency loudspeaker of the three-way dodecahedron loudspeaker developed by ITA was used. The sweep was chosen to comply with Eq. 3.19 and hence no overlapping occurs. Therefore, only errors due to level changes in the fundamental are expected. Figure 3.30 shows the observed dependence of the absolute value of sound strength for the 4 kHz third octave band due to a change of amplification level. The actual change of the level is compensated for during the measurement. Hence, the observed deviation is due to a nonlinear element in the measurement chain, which is assumed to be the loudspeaker. The scale for the output amplification is not directly comparable to the one in the results shown before. THD was below 10 % for -10 dBFS in this measurement. The bars in the graph correspond to the standard deviation observed over several measurements with the same output amplification at the same position.

As can be seen, the deviation from the sound strength determined at low output amplifications and hence low THD increases with increasing output amplification. However, the sign of the deviation is opposite to the simulated deviations. This can be explained as the input-output-diagram generated by odd order polynomial does not directly apply for electro-dynamic loudspeakers. The input-output-diagram of a loudspeaker in general shows a limiting behavior and might be approximated by $y = x - x^3$ instead. Hence, the sign of the deviations simulated before should be inverted to be compared with the measurement.

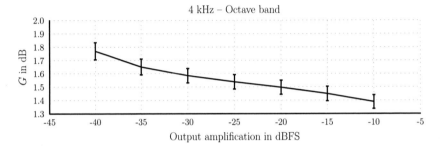

Figure 3.30.: Measured dependence of the sound strength due to a change of amplification level in an auditorium with a dodecahedron loudspeaker driven in a slightly nonlinear range.

3.8. Summary and Scientific Contribution

Uncertainties in the impulse response and their possible sources have been investigated by simplifying and describing the entire measurement chain with a black box approach. This approach also leads to a calibration method that generally enables the measurement of absolute transfer functions.

The focus was set on the modeling and simulation of nonlinear systems to study the principal influence on the measured impulse responses. The characterization of nonlinear systems has been presented. It was enhanced by especially deducing the relationship between the polynomial coefficients of a simple nonlinear system with the harmonic impulse responses observable in exponential sweep measurements. The variation of the level of the input signal on the amplitude of the harmonics and the phase of these harmonics in relation to the fundamental can hence be predicted. Moreover, the influence of distortion on the fundamental impulse response was analyzed in detail. It has been shown that the fundamental impulse response in measurements with exponential sweeps is not identical to the linear impulse response of the system and that applying a time window to cut harmonic impulse responses does not necessarily suppress the distortion artifacts entirely. Hence, the level of the harmonics should be continuously monitored during measurements to ensure sufficiently low distortion and hence low uncertainties in the obtained transfer functions.

The original idea to characterize the system with nonlinear elements in the measurement chain using a nonlinear model and to calculate the linear response in a post-processing step has been discarded. This approach is not applicable for typical acoustical measurements chains where the nonlinear element has to be modeled by a Wiener-Hammerstein model followed by the impulse response of the mostly linear DUT. However, the findings from the nonlinear theory were applied yielding a method to suppress certain harmonics by superposing multiple sweep measurements with different sweep parameters.

Besides the harmonic impulse responses usually observed in exponential sweep measurements of weakly nonlinear system, additional impulse responses due to inter-modulation might occur. This is the case if the combined impulse response of the elements before the nonlinear element deviates from an ideal Dirac function. The relation between these inter-modulation impulse responses and the parameters of the exponential sweep have been deduced. These equations are then used to optimize the sweep parameters and hence allow to generally avoid such artifacts.

Quantization artifacts are usually considered as white noise in the measurement. In case a very high SNR is desired or the amplitude of the signal is not far above the quantization noise, dithering of the excitation signal in conjunction with averaging can significantly improve the SNR and minimize distortion.

The modeling of these artifacts, e.g., quantization, noise and nonlinearities, and the measurement procedure itself along with the linear transfer function is summarized in a simulation model of a typical measurement chain in acoustics. It was implemented in MATLAB for the ITA-Toolbox and made public (DIETRICH et al., 2013; DIETRICH, GUSKI, and VORLÄNDER, 2013). An example from room acoustics has been presented to analyze the principal uncertainties in the room acoustic parameters caused by nonlinearities in the loudspeaker. The uncertainties in the room acoustic parameters caused by modeled nonlinearities of a loudspeaker have been found to be potentially in the same order of magnitude than the just noticeable difference of the parameters. Two different effects have been investigated. Firstly, overlapping of harmonic impulse responses with the fundamental impulse response, that is used to evaluate the room acoustic parameters. Secondly, deviation in the fundamental compared to the linear impulse response. However, the applications are manifold and this approach has proven to be very powerful for the detailed investigation of measurement uncertainties introduced by elements of the measurement chain for specific measurement tasks and equipment.

4

Uncertainties in Airborne Transfer Paths

This chapter concentrates on modeling and analyzing airborne scenarios with one sound source and one or more receivers in the acoustic space. A characterization of sources and receivers mainly in terms of their directivity or radiation pattern as well as the linear transfer function between these two points are introduced. Parametric models for three exemplary applications are presented—one from a measurement of a sound barrier and two from room acoustics. Finally, the uncertainties of derived quantities are investigated using Monte Carlo simulations in combination with the uncertainty of the input quantities describing the positioning accuracy of the source or the receiver.

4.1. Modeling Sound Sources and Receiver

A sound source as required for acoustic measurements can be considered as a black box that radiates sound into the surrounding medium air. In general, this black box is in interaction with the medium. But unlike in structure-borne sound this coupling effect is rather small and can be neglected in most practical cases as also done in this thesis. The only coupling considered is described by the radiation impedance into the free-field. The radiation pattern is in general frequency dependent. Simple sound sources internally consist of a single signal source and one radiation pattern. Even multipath loudspeakers could be described in such a way regardless of their multiple membranes or drivers with different frequency responses if they are fed with correlated or even the same signals. In case the multiple ways are fed with uncorrelated or multiple signals a radiation pattern for each way has to be considered to capture the correct radiation. Musical instruments or complex sound sources, e.g., combustion engines or household appliances, cannot in general be described in such a simple manner as they might consist of multiple independent (uncorrelated) signal sources associated with different radiation patterns. However, even for musical instruments such assumptions are made to reduce the complexity to a reasonable degree and the synthesized results still sound plausible (VORLÄNDER, 2007).

Directivity patterns of simple sound sources can directly be expressed analytically. Simple basic patterns are the *monopole* (omni-directional), *dipole* and *quadrupole* (MECHEL, 2008; WILLIAMS, 1999) which are used in the following examples. Since reciprocity theorem holds in linear acoustics the same characterization used for sound sources can be directly applied to receivers. Therefore, the terms sources and receivers can be understood synonymously in the following.

4.1.1. Point Source in Free-Field

The radiation of an ideal omni-directional source can be expressed as

$$G_0\left(r\right) = \frac{p(\omega, r)}{\Psi(\omega)} = \frac{e^{jkr}}{4\pi r} \qquad (4.1)$$

where $G_0(r)$ is the Green's function with dimension $1/\text{m}$ for the monopole and r the distance between source and receiving position at which the sound pressure is measured. Ψ is introduced as the airborne source descriptor for the monopole. The source factor for a vibrating sphere with the radius of r_{sphere} and the normal velocity v_{mem} can be given as

$$\Psi = j\omega \cdot \rho_0 \frac{4}{3}\pi r_{\text{sphere}}^2 \cdot v_{\text{mem}} \qquad (4.2)$$

where ρ_0 is the density in air.

4.1.2. Directivity Patterns—Spherical Harmonics

Spherical Harmonics (SH) provide a description for directivity patterns using *spherical base functions* and associated *SH coefficients*. They use the concept of a two-dimensional Fourier transformation on a sphere (WILLIAMS, 1999)[1].

---

[1] Parts of the following are already published in (POLLOW et al., 2012).

SH can be used to describe any two-dimensional square-integrable spatial function $f(\theta, \phi)$ as a set of spherical harmonic coefficients noted as $\hat{a}_n^m$ with the order n and the degree of the coefficients m. Spherical functions depend only on the angles θ and ϕ (angles depicted in Figure 4.1 on the left side). The complex SH *base functions* $Y_n^m(\Omega)$ are defined as

$$Y_n^m(\theta, \phi) = \sqrt{\frac{2n+1}{4\pi} \frac{(n-m)!}{(n+m)!}} \, P_n^m(\cos\theta) \, e^{jm\phi} \tag{4.3}$$

where P_n^m is the *Legendre function* of the first kind of order n and degree m.

Directivities for sound sources and receivers can be expressed by their frequency dependent spherical harmonics coefficients as, e.g., shown in (POLLOW, 2007). Directivity measurements can be transformed to this representation as exemplarily illustrated in Figure 4.1 for a cuboid loudspeaker enclosure with a single membrane for one particular frequency around 1 kHz. This representation is similar to the Fourier transform between time and frequency domain. Details on the transformation and the spatial discretization are given in (ZOTTER, 2009).

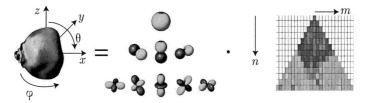

Figure 4.1.: Exemplary relation between a loudspeaker radiation pattern (left) and superposition of real spherical harmonic base (center) functions with SH coefficients (right) for a specific frequency [after (KUNKEMÖLLER, 2011)]. The color in the spherical plots represents the phase (red: 0° blue: 180°).

4.1.3. Loudspeaker Cap Model

To provide a realistic but yet analytic model for a loudspeaker consisting of a membrane mounted in a spherical housing as a sound source with directivity the so-called *cap model* can be used. It assumes a sphere with radius r_{sphere} with a vibrating membrane modeled as a cap with radius r_{mem} with a velocity v_{mem}. Figure 4.2 shows a north pole cap with the aperture angle α defined as

$$\alpha = \arcsin\left(\frac{r_{\text{mem}}}{r_{\text{sphere}}}\right). \tag{4.4}$$

The analytic model is described in (POLLOW, 2007; POLLOW and BEHLER, 2009) and uses the SH representation for north pole caps. These are then rotated to arbitrary angles. The *Wigner-D* rotation matrix is used for the implementation of the rotation. The north pole cap is rotationally symmetric around the central axis of the sphere and can be described by spherical harmonic coefficients of the degree $m = 0$. The SH coefficients of the north pole cap can be calculated as

$$\hat{a}_{n,\mathrm{cap}}^0 = \sqrt{\pi(2n+1)} \cdot \int_{\cos\alpha}^1 P_n(x)\mathrm{d}x = \frac{P_{n-1}\left(\cos\alpha\right) - P_{n+1}\left(\cos\alpha\right)}{2n+1}. \quad (4.5)$$

Details on the implementation of the model used and the rotation to arbitrary angle can be found in (J. KLEIN, 2012). The radiated sound pressure in SH at an arbitrary distance r is given by

$$p_n^m(r) = \frac{h_n(k \cdot r)}{h_n(k \cdot r_{\mathrm{sphere}})} \cdot p_n^m(r_{\mathrm{sphere}}). \quad (4.6)$$

where h_n is the *Hankel function* of the second kind and the wave number $k = 2\pi f/c$. The *kr-limit* can be used as a *rule of thumb* to approximate the highest SH order n_{max} to be considered with the maximum frequency f_{max} of interest according to (DURAISWAMI, ZOTKIN, and GUMEROV, 2004)[2] as

$$n_{\mathrm{max}} = \lfloor k_{\mathrm{max}} \cdot r_{\mathrm{sphere}} + 1 \rfloor = \left\lfloor \frac{2\pi \cdot f_{\mathrm{max}}}{c_0} \cdot r_{\mathrm{sphere}} + 1 \right\rfloor. \quad (4.7)$$

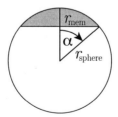

Figure 4.2.: Spherical cap model with sphere radius r_{sphere} and membrane radius r_{mem} building an aperture angle α used for the analytic model to model the directivity of a spherical loudspeaker.

---

[2] A constant offset of $+1$ is used here to safely include the maximum frequency.

4.2. Application I—Reflection Index of Sound Barriers

The *work package 3* of the QUIetening the Environment for a Sustainable Surface Transport (QUIESST) project[3] focuses on the measurement of sound reflection properties of sound barriers. Especially the constant characterization and therefore the monitoring of sound barriers for highways or railroads is under research. It is based on the standard (CEN/TS 1793-5, 2003). Seven laboratories participated in round robin measurements at two test sites—one in Grenoble, France and one in Valladolid, Spain—in 2011. These results are supposed to be used to determine the overall measurement uncertainty by comparing results between the different laboratories and to obtain a standard deviation.

The intention in this thesis is to derive uncertainties by using an acoustic model for the transfer functions and hence follow more closely the Guide to the expression of Uncertainty in Measurement (GUM) guideline and express significant uncertainty contributions already found in the practical measurements separately. This is not possible by analyzing the final results of the different laboratories only. Furthermore, systematic errors can be observed more easily based on simulation data, as the ideal characteristics of the barrier are defined manually and hence they are known in contrary to experiments in the field.

4.2.1. Measurement Method for Reflection Index

The QUIESST procedure consists of two measurements of transfer functions and one post-processing step to derive the Reflection Index (RI) as the measurement quantity. To obtain a calibrated measurement the transfer functions between the loudspeaker and the 9 pressure microphones have to be obtained in two different conditions. Firstly, in the acoustic free-field and secondly, in front of the sound barrier as the Device Under Test (DUT). Details on positioning, exact geometry and further post-processing can be found in, e.g., (CLAIRBOIS et al., 2012).

The post-processing uses the measured impulse responses and the *ADRIENNE* time window technique to fade out unwanted reflection components, e.g., from the ground and other objects as long as they appear later than the direct and the reflected component. The specified length of the time window is 7.9 ms. Direct

---

[3]The Institute of Technical Acoustics at RWTH Aachen University (ITA) took part in the *work package 3* of this project co-funded by the European Community's Seventh Framework Programme (FP7/2007-2013) (www.quiesst.eu/).

and reflected component cannot be directly separated by a time window. As the impulse response of the loudspeaker is usually longer than this delay a shorter time window would introduce a frequency smoothing effect for the low frequency (approx. $< 120\,\text{Hz}$) (MÜLLER, 1999). However, equalization of the loudspeaker's impulse response can reduce the observed impulse responses for the direct sound and the reflected component, e.g., (WEHR et al., 2013).

The definition of the *reflection index* RI for each microphone l is given as

$$\text{RI}_{\text{mic}_l} = \frac{\int\limits_{\Delta f_j} |\mathcal{F}\{h_{\text{r},l}(t) \cdot w_{\text{r},l}(t)\}|^2 \, \text{d}f}{\int\limits_{\Delta f_j} |\mathcal{F}\{h_{\text{d},l}(t) \cdot w_{\text{d},l}(t)\}|^2 \, \text{d}f} \cdot c_{\text{dir}}(f_j) \cdot c_{\text{geo}} \qquad (4.8)$$

where f_j is the mid frequency of the one-third octave band j (GARAI, 2011). Furthermore, h_{d} is the direct sound and w_{d} the corresponding time window, h_{r} and w_{r} are used for the reflected component.

A frequency dependent correction factor $c_{\text{dir}}(f_j)$ accounts for the deviation from the omni-directional radiation pattern of the loudspeaker. The remaining correction factor c_{geo} accounts for the traveling distance between direct sound and the reflected component based on a point source assumption. Both factors can be omitted if a second reference measurement with a distance of 175 cm between the loudspeaker and the microphone array is used to obtain the $h_{\text{r},l}(t)$ used in the denominator in Eq. 4.8. The method involving the additional measurement is used in the following. Possible values of the reflection index range from 0 according to absolutely no reflection by the barrier to 1 for a totally reflecting barrier. The values are always real-valued and positive. However, in practice values greater than 1 can be observed for non-flat but highly reflective surfaces for specific microphone positions.

The reflected component h_{r} is obtained by a measurement in front of the barrier h_{barrier} and subtracting the direct component obtained in the free field as

$$h_{\text{r}}(t) = h_{\text{barrier}}(t) - c_{\text{equal}} \cdot h_{\text{d}}\left(t - \frac{\Delta n_{\text{min}}}{f_{\text{s}}}\right). \qquad (4.9)$$

An additional sub-sample shift in the range of $\Delta n = \pm 2$ samples is specified to align the two direct components prior to subtraction. For a sampling rate of $fs = 44100\,\text{kHz}$ these maximum shifts correspond to a maximum radial positioning error of ± 1.56 cm that could be compensated assuming $c = 344$ m/s[4]. An algorithm is specified to obtain $\Delta n_{\min}$ such as the residue with ± 50 samples around the main peak position of the direct sound is minimized. Furthermore, the direct sound is equalized with a factor c_{equal} such as the main peaks in both impulse responses are equalized prior to the subtraction[5]. The time windows are placed around the peaks of the direct sound and the reflected component, respectively.

The loudspeaker has to radiate low frequencies starting below 100 Hz and up to at least 8 kHz. The radiation pattern was initially assumed to be omni-directional. This was discarded during the project phase and hence, the correction factor $c_{\mathrm{dir}}(f_j)$ was introduced. Crossover networks, multiple membranes and substitution of the loudspeaker for different frequency ranges is not permitted by the new procedure. Hence, a loudspeaker for this purpose was developed by the ITA using only a single membrane. For the low frequencies the membrane has to be of a certain size to allow enough sound pressure level without leaving a linear range. This yields in loudspeaker with a prominent directional pattern at the highest frequencies of interest. Diffraction effects were minimized by choosing an enclosure without sharp edges as shown in Figure 4.4 and similar to the analytic model shown in Figure 4.2.

The center of the loudspeaker membrane has to be 1.25 m in front of the center of the 3×3 planar microphone array[6]. For the reference measurement this distance has to be increased to 175 cm The microphones are spaced by 40 cm in horizontal and vertical direction and the array is then placed parallel to the sound barrier with a distance of 25 cm. The developed measurement setup is shown in Figure 4.4 during a free-field calibration measurement of $h_{\mathrm{d},l}(t)$. The microphone positions can be divided into three classes as shown in Figure 4.3.

---

[4]The current formulation of the measurement procedure does not specify the maximum frequency range or the sampling rate to be used. Hence, this time shift becomes smaller with increasing sampling rate.

[5]This equalization method was discussed and criticized as this could result in a systematic error that could be avoided if h_{barrier} was scaled with the inverse factor instead.

[6]The array might also consist of 9 microphones placed consecutively. The measurement method currently also suggests to avoid a fixed connection to the loudspeaker. However, the measurement setup shown involves a rigid connection minimizing position error and the setup time.

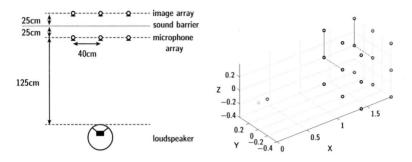

Figure 4.3.: Geometry of the measurement setup with dimensions in meters, top view (left) and 3-D view (right) with center of spherical loudspeaker (red), membrane (red, bold), microphone array (black), image array (green) and the three classes of microphone positions for each array (blue).

4.2.2. Modeling of the Measurement Setup

Two main sources of uncertainty have been observed empirically in the field. Firstly, the relative position of the microphone to the middle of the loudspeaker. Secondly, the orientation of the loudspeaker in conjunction with its directivity. Further uncertainties are introduced due to nonlinearities of the loudspeakers as described in Section 3.3 and especially Figure 3.7 for the room acoustic example due to both overlapping and changes in level between reference and in-situ measurement. The decay of loudspeaker's impulse response has an impact in conjunction with the length of time window applied. This uncertainty can also be studied with the presented model but is not shown, as the impulse response of the loudspeaker used was considered sufficiently short. Furthermore, wind might be a source of uncertainty if present as wind can introduce uncorrelated noise originated at the microphones or sharp edges in the vicinity of the measurement setup and it might change the transfer characteristics in air. This is not considered in this thesis as the problem is analyzed in detail in (X. WANG, (2013)) for the same measurement setup.

The microphones are modeled as ideal omni-directional receivers, neglecting diffraction at the microphone cylinders, cables, the array and the microphone stands.

The loudspeaker is modeled according to the cap model with $r_{\mathrm{mem}} = 6.75\,\mathrm{cm}$ and $r_{\mathrm{sphere}} = 15\,\mathrm{cm}$ and with a maximum SH order of $n_{\max} = 23$ corresponding

Figure 4.4.: Array with 18 (9 array microphones and shifted array) microphones and spherical loudspeaker in a free-field reference measurement during the QUIESST Round Robin in Valladolid, Spain.

to a maximum frequency of $f_{\max} = 8.3\,\mathrm{kHz}$ according to Eq. 4.7. The simulated directivity pattern is depicted in Figure 4.5. The sound barrier is modeled analytically by a locally reacting plain wall with infinite dimensions. Hence, edge diffraction at the top and the sites of the barrier is neglected. The angle-independent impedance of the barrier Z_{barrier} is calculated based on a mean reflection factor R_{mean} used to vary the reflectivity of the simulation model as (KUTTRUFF, 2007)

$$Z_{\mathrm{barrier}} = \frac{1 + R_{\mathrm{mean}}}{1 - R_{\mathrm{mean}}} \frac{Z_0}{\cos(\theta)} \tag{4.10}$$

with θ the angle between the sound incidence at the wall and the wall normal vector and the free-field impedance in air $Z_0 = \rho_0 c$. Hence, the impedance is always real-valued in the simulations. The angle-dependent reflection factor of a locally reacting wall assuming plane waves is expressed as

$$R\left(\theta\right) = \frac{\cos(\theta) - \frac{Z_0}{Z}}{\cos(\theta) + \frac{Z_0}{Z}}. \tag{4.11}$$

This effect of angle dependence is stronger for low values of R_{mean} and hence, low values of Z_{barrier}. For high reflection factors the wall impedance goes to infinity and the effect vanishes.

The reflected component is calculated assuming an image microphone array (25 cm behind the frontal surface of the barrier and hence without considering the thickness of the barrier) with the distance of 175 cm to the loudspeaker. Additionally, the angle-dependent attenuation due to the reflection in Eq. 4.10

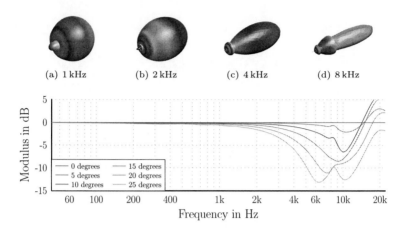

Figure 4.5.: Directivity patterns for different frequencies in linear scale (top row) and frequency response normalized to *on-axis* direction $0°$ for different angles obtained with the spherical cap model ($r_{mem} = 6.75\,cm$, $r_{sphere} = 15\,cm$) for $n_{max} = 23$ and $f_{max} = 8.3\,kHz$. The center of the membrane is used as rotational fix point.

is accounted for. Figure 4.3 illustrates the positions of the loudspeaker and the microphones. The ground is not modeled to avoid artifacts due to interferences with the reflected components in $h_{r,l}(t)$ and to concentrate of the primary uncertainties due to the positioning only.

The following simulations provide more insight into the causes of the uncertainty in measurements of RI. Although these results are gained for ideal—flat and infinitely large—barriers, the findings are assumed to be applicable to real-world scenarios.

Uncertainty—Microphone Position

The uncertainty for the microphone positions is modeled by three normal distributions around the exact microphone positions for the x, y- and z-direction, respectively. The standard deviation is denoted as u_{mic}. Based on an empirical analysis in the field and in the laboratory the uncertainty in positioning is estimated to lie between 2.5 cm and 7.5 cm. This range depends on whether a fixed array of microphones or single microphone measurements is used, and whether the microphones are fixed to the entire setup, especially to the loudspeaker.

This fixation increases the accuracy of the positioning. Furthermore, it depends on the conditions in the field, i.e. flatness and softness of the ground, where the loudspeaker and microphone stands are to be placed. Figure 4.6 shows the modeled measurement setup. Three different measurements are simulated—a reference measurement in 125 cm, a reference measurement in 175 cm and the in-situ measurement setup in front of the barrier including the attenuation by the angle-dependent reflection factor. Monte Carlo (MC) simulations with 100 runs are carried out for each of the three measurement setups, where the microphone positions are perturbed for each microphone without correlation. This represents the case when no fixed connection between the microphones is provided. For the in-situ measurement the microphone positions of the array are varied and the position of the image array representing the receiving microphones follow mirrored at the barrier.

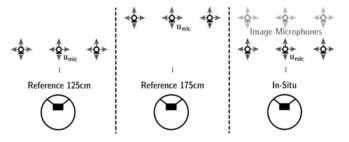

Figure 4.6.: Simulation setup to model the uncertainty in microphone position for the two reference measurements in 125 cm and 175 cm distance and the in-situ measurement in front of the barrier where the image microphones are directly linked to the position of the array microphones.

The simulations include the loudspeaker's directivity as the cap model is used. Figure 4.7 shows the variation of the impulse response in front of the barrier with a mean reflection factor of $R_{\mathrm{mean}} = 0.7$. In the time domain the direct sound and the reflected component vary both in amplitude and arrival time. In the frequency domain the comb-filter changes accordingly depending mainly on the time delay between both peaks. The distribution of the microphone positions is exemplarily shown in Figure 4.8 for a standard deviation of $u_{\mathrm{mic}} = 2.5$ cm.

The results for the reflection index are shown in Figure 4.9 for $u_{\mathrm{mic}} = 2.5$ cm and $u_{\mathrm{mic}} = 7.5$ cm. They are divided into the three classes of microphones and the averaged reflection factor RI, which is the weighted mean of the results from each microphone position. The standard deviation of RI_l increases with increasing

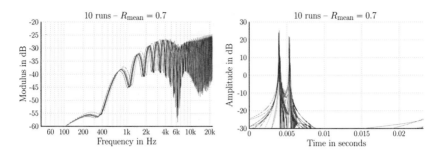

Figure 4.7.: Impulse response (left) in front of the barrier and corresponding frequency response (right) for 10 runs with $u_{mic} = 2.5\,cm$ and $R_{mean} = 0.7$ using the analytic model.

uncertainty in the positioning (left column compared to the right column). The center microphone (Mic: 5) shows the smallest deviations, whereas the microphone with the largest angle in relation to the main axis of the loudspeaker (Mic: 3) shows the highest deviation. With increasing angle the radiation pattern becomes more prominent and influences the sensitivity of the RI_l to positioning errors. The mean reflection index RI shows lower uncertainty than the single contributions. This is due to an averaging effect over the 9 microphone positions. Especially, as the perturbations in the microphone positions are uncorrelated.

With increasing frequency the uncertainty is expected to increase due to the radiation pattern. This effect can be observed for simulations of RI_l and the mean RI. The uncertainty (Figure 4.9, bottom row) scales with the reflection factor of the wall. Hence, highly reflective barriers can only be measured with higher uncertainties for the upper frequency bands.

It is important to investigate the larger deviations at the third-octave bands below 400 Hz for $u_{mic} = 7.5\,cm$. At these frequencies the radiation pattern is very close to the omni-directional pattern. Hence, no large deviations are expected, except for the deviations due to positioning errors radial to the loudspeaker. During the measurements in the field, these deviations where assumed to be caused by an Signal-to-Noise Ratio (SNR) due to limited sound pressure produced by the real loudspeaker for this frequency range. However, the simulations do not include any additive noise sources. Due to these simulations, this effect was found to be caused by the adaptive time windowing in the QUIESST procedure. The radiation pattern effects mainly higher frequencies for the off-axis microphone positions. The variation of these positions yields large differences in level compared to the

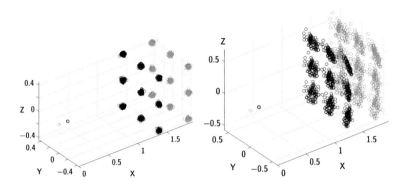

Figure 4.8.: Distribution of the relative position between microphones and loudspeaker for uncertainty in microphone position with $u_{\mathrm{mic}} = 2.5\,\mathrm{cm}$ (left) and uncertainty in loudspeaker orientation $u_{\mathrm{ls}} = 2.5°$ with 100 runs each.

lower frequencies. The peak of the impulse responses strongly depends on the high frequency content. The adaptive time window routine is very sensitive to changes of these peaks. The minimum residue during the automatic subtraction algorithm is found for an incorrect time shift as it does not reflect the deviation in traveling time. The incorrect time alignment results in an increased residue for low frequencies. Hence, the deviation in RI_l is increased for these low frequencies caused by an overlapping effect similar to room acoustic example with nonlinearities in Section 3.7.2. This overlapping of the residue of the direct sound has a stronger influence on the result if the energy of the reflection is low. Hence, the uncertainty in RI for low frequency bands potentially increases with decreasing reflection factor of the barrier.

In order to prove the statement regarding the subtraction method, the same simulation results but without using the subtraction method are calculated. This is only possible by means of simulations. The reflected component as already simulated is directly used as input in Eq. 4.8. The results are shown in Figure 4.10 in the same manner as before. Especially the difference in the standard deviations can be used as a benchmark for the subtraction method. For $u_{\mathrm{mic}} = 2.5\,\mathrm{cm}$ both results are almost identical but for $u_{\mathrm{mic}} = 7.5\,\mathrm{cm}$ the results show that the uncertainties for the lowest frequency bands are caused by subtraction method. The straight line for the low frequencies further prove that the uncertainty in this region is mainly caused by the level difference caused by deviations in position perpendicular to the loudspeaker. As a consequence, the evaluation method could benefit from a segmentation of the procedure into different frequency bands. E.g.,

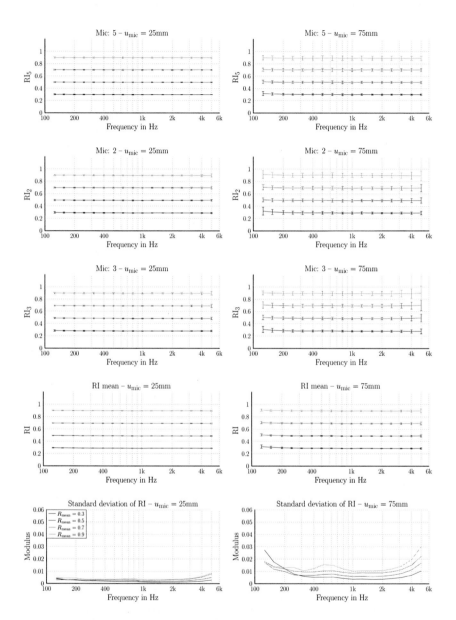

Figure 4.9.: Reflection index for three single microphones (row: 1,2 and 3) and mean reflection index RI (bottom row) with standard deviation obtained by Monte Carlo simulations with $u_{mic} = 2.5\,cm$ (left) and $u_{mic} = 7.5\,cm$ (right) for different reflection factors R_{mean}.

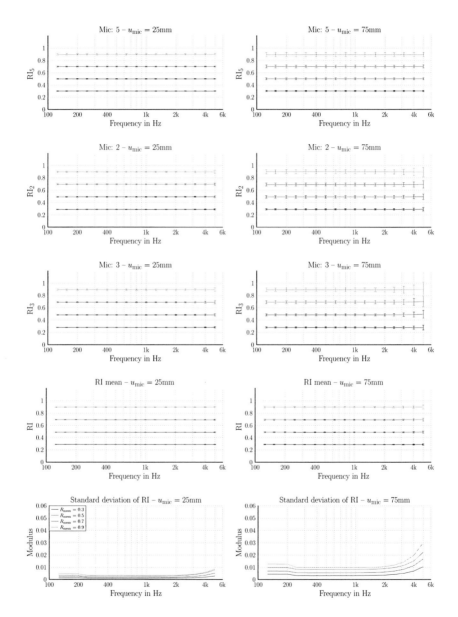

Figure 4.10.: Reference reflection index calculated directly from the simulated reflected component and hence without using the subtraction method.

low-passed impulse responses could be used to obtain the low frequency range and the broadband or high-passed data could be used for the upper frequency bands to minimize further uncertainties caused by the subtraction.

For $u_{\mathrm{mic}} = 7.5\,\mathrm{cm}$ the uncertainty for the lowest frequency bands is already dominated by the overlapping effect and hence increases with decreasing reflection factors. Investigations of the search range (standard is ± 2 samples) for the automatic subtraction showed variation of the observed uncertainty. However, it cannot be stated that either a fixed subtraction or an adapted search range taking into account the actual delay caused by a shift of 7.5 cm decreases the uncertainty for all frequencies of interest.

It is important to point out that the mean of the averaged RI matches the given R_{mean} for $u_{\mathrm{mic}} = 2.5\,\mathrm{cm}$ except for the small deviation expected for lower reflection factors due to the angle dependence. For $u_{\mathrm{mic}} = 7.5\,\mathrm{cm}$ only the mean of the two lowest frequency bands for $R_{\mathrm{mean}} = 0.3$ deviate.

The uncertainty in the reflection index can be significantly reduced if the precision in positioning the microphones relative to the loudspeaker can be improved. Especially the impact of the overlapping effect due to a mismatched subtraction for low reflection factors is considered problematic as this is not the only uncertainty contribution.

Uncertainty—Loudspeaker Orientation

The modeled measurement setup is shown in Figure 4.11 similar to the one used for the position uncertainty. MC simulations are carried out for the three measurements with the same number of runs. The orientation angle of the loudspeaker is chosen from a normal distribution with standard deviation u_{ls}. Depending on the conditions in the field this uncertainty is estimated to lie between $1°$ and $5°$ based on empiric data. The distribution of the microphone positions is shown in Figure 4.8 for a standard deviation of $2.5°$. The rotation of the loudspeaker is compensated in the plot. Hence, the microphone positions vary instead.

The simulation results are depicted in Figure 4.12 for the three classes of microphones, the mean RI and the standard deviation. In comparison to the results with uncertain microphone positions the deviation for the low frequency bands are smaller for the lowest frequency bands. This is although the spread of the

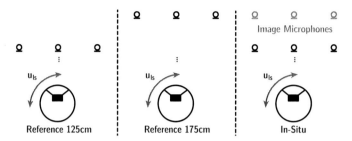

Figure 4.11.: Simulation setup to model the uncertainty in loudspeaker orientation for the two reference measurements in 125 cm and 175 cm and the in-situ measurement in front of the barrier.

microphone positions is larger for exemplarily chosen values for the loudspeaker orientation uncertainty than for the microphone position uncertainty as shown in Figure 4.8.

As the rotation of the loudspeaker does not influence the traveling time for the direct sound component, the automatic subtraction works almost perfectly. With increasing frequency the deviations grow as the radiation pattern of the loudspeaker becomes prominent. The uncertainty of RI increases with increasing uncertainty of the loudspeaker orientation as expected. The off-axis microphone positions show larger deviations especially at high frequencies but the mean RI has a considerably low uncertainty compared to the single microphones. Although the microphone positions relative to the loudspeaker's main axis are perturbed with a strong correlation, the mean RI provides still an averaging effect and the uncertainty is significantly decreased compared to the RI_l. For $u_{ls} = 2.5°$ the standard deviation is comparable to the results obtained for an uncertainty in position of the microphones with $u_{mic} = 2.5$ cm except for frequencies above approx. 2 kHz for the chosen loudspeaker directivity. The uncertainty also increases with increasing reflection factors.

For $u_{ls} = 7.5°$ the standard deviation in RI is generally higher for all frequencies and all reflection factors. This can be explained again by the overlapping effect. However a strong dependence on the reflection factor cannot be observed. Moreover, the mean of $RI_|$ does also not deviate from the ideal values. Especially for large uncertainties in the orientation of the sound source it is important to conduct a reference measurement each time a repeated measurement should be conducted. Otherwise, a reference with a strong deviation for higher frequencies will be used for all measurements and the mean of the repeated measurements

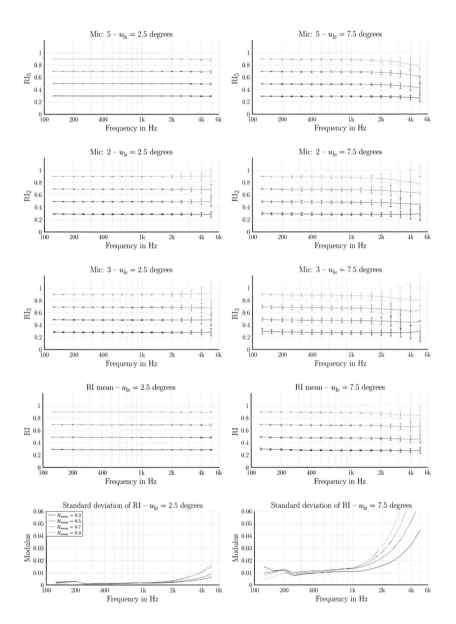

Figure 4.12.: Reflection index for three single microphones (row: 1,2 and 3) and mean reflection index RI (bottom row) with uncertainty in loudspeaker orientation $u_{ls} = 2.5°$ (left) and $u_{ls} = 7.5°$ (right).

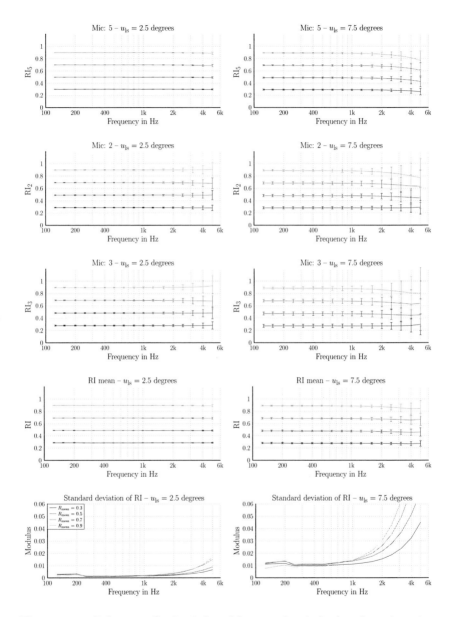

Figure 4.13.: Reference reflection index with uncertainty in loudspeaker orientation but directly using the simulated reflected component and hence without using the subtraction method.

will show a drift to either higher or lower values for the high frequency bands. The Monte Carlo simulations shown used a different reference measurement for each run. The ideal results without using the subtraction method are depicted in Figure 4.13 to allow a comparison.

For larger uncertainties in positioning which are assumed to be more realistic under non ideal conditions in the field, the combined uncertainty is calculated to lie below ±0.03 for all reflection factors up to a frequency of approx. 2 kHz. Above this frequency, the uncertainty increases further and it is dependent on the reflection factor of the barrier.

The simulated uncertainties are all based on an approach that emulates repeated measurements with the same equipment. Noticeable deviations of the mean value have not been found in the result. Otherwise, these deviations have to be included in the uncertainty budget. The directivity of the loudspeaker was chosen based on real source available for measurements. As long as the measurement procedure does not specify the directivity pattern or its constraints any type of loudspeaker with a single membrane might be used. The differences between results of different teams with different equipment and a similar positioning accuracy are always larger than the ones obtained by the presented procedure. Other sources of uncertainty, e.g. meteorological condition and background noise, might become dominant if the positioning accuracy was improved. Results from the round robin in the QUIESST project are not shown in this thesis, but the deviations are commonly larger than the simulated results. This is assumed to be additionally caused by different implementations of the evaluation procedure as a project internal software round robin with the same input data supported this assumption and different measurement equipment used.

4.3. Room Modes and Modal Superposition

The previous section dealt with sound propagation in the free-field with only one reflecting surface. Enclosures with six reflecting surfaces are studied in the following leading to the field of room acoustics. An analytic model for weakly damped rectangular rooms from the literature is implemented and is enhanced towards directional sources and receivers[7]. The relation to measurement data is given by applying the ration fit on measurement data. Room acoustic parameters

---

[7] Parts of following sections have been partly published in (POLLOW, DIETRICH, and VORLÄNDER, 2013)

are derived to show the applicability of the approach. This analytic model is later used to study perturbations in the position of the sensors and perturbations in the orientation of the directional sound sources.

4.3.1. Analytic Model for Rectangular Rooms

Modal superposition for rectangular rooms with rigid boundaries is described, e.g., in (KUTTRUFF, 2000; MECHEL, 2008; PIERCE, 1989). Analytic formulations for non-rigid walls have been published, e.g., in (BISTAFA and MORRISSEY, 2003; NAKA, OBERAI, and SHINN-CUNNINGHAM, 2005). Rigid walls are assumed in the following. The eigenfrequencies for the room of dimensions $L_x \times L_y \times L_z$ are given as

$$\omega_i = c\pi \sqrt{\left(\frac{n_{x,i}}{L_x}\right)^2 + \left(\frac{n_{y,i}}{L_y}\right)^2 + \left(\frac{n_{z,i}}{L_z}\right)^2} \tag{4.12}$$

with the modal numbers $n_{x,i}$, $n_{y,i}$ and $n_{z,i}$ being all possible combinations of non-negative integer values and c the speed of sound. The origin of the coordinate system is set to be in the corner. The room transfer function from a point source at $\mathbf{r}_s$ to a receiver point at $\mathbf{r}_r$ is obtained by superposition in the frequency domain as

$$H(\omega) = -\frac{4\pi c^2}{V} \sum_i \frac{\psi_i(\mathbf{r}_s)\psi_i(\mathbf{r}_r)}{(\omega^2 - \omega_i^2 - \mathrm{j}\delta_i\omega_i)K_i} \tag{4.13}$$

with the volume $V = L_x \cdot L_y \cdot L_z$, a modal damping constant δ_i and $K_i = \iiint \psi_i^2(\mathbf{r})\,\mathrm{d}V$ a normalization constant for the eigenmodes. The abstract modal damping constant is related to the more practical modal reverberation time RT_i by

$$\delta_i = \frac{3 \cdot \ln(10)}{\mathrm{RT}_i}. \tag{4.14}$$

The modal reverberation time has to fulfill (KUTTRUFF, 2000)[8].

$$\mathrm{RT}_i \gg \frac{3\ln(10)}{2\pi f_i}. \tag{4.15}$$

The eigenfunction in the room sampled at the position $\mathbf{r} = (x, y, z)$ reads as

$$\psi_i(\mathbf{r}) = \cos\left(\pi n_{x,i}\frac{x}{L_x}\right) \cos\left(\pi n_{y,i}\frac{y}{L_y}\right) \cos\left(\pi n_{z,i}\frac{z}{L_z}\right). \tag{4.16}$$

---

[8]Assuming a lowest frequency of 10 Hz the reverberation time has to be greater than 0.1 s

95

For each room there exists an infinite number of modes, while in practice an upper frequency limit is used to restrict the number of modes for calculation. The number of modes N up to an upper frequency f_{max} can be approximated according to (KUTTRUFF, 2000) as

$$N(f_{max}) = \frac{4\pi}{3} V \left(\frac{f_{max}}{c}\right)^3 + \frac{\pi}{4} S \left(\frac{f}{c}\right)^2 + \frac{L}{8} \left(\frac{f}{c}\right) \approx \frac{4\pi}{3} V \left(\frac{f_{max}}{c}\right)^3 \quad (4.17)$$

with $V = L_x L_y L_z$, $S = 2\left(L_x L_y + L_y L_z + L_y L_z\right)$ and $L = 4\left(L_x + L_y + L_z\right)$. Hence, the modal density dN/df increases mainly quadratically over frequency. The three terms stand in order of appearance for the number of 3-D, 2-D and 1-D modes respectively. 1-D means that only one of the three n is different from zero and 2-D means two differ from zero, respectively. As an example, a room with dimensions $8 \times 5 \times 3$m has approx. $1.2 \cdot 10^8$ eigenfrequencies for $f_{max} = 20$ kHz. Therefore, this analytic model can only be applied with reasonable computation times for either small rooms (compared to the wavelength) or a limited frequency range.

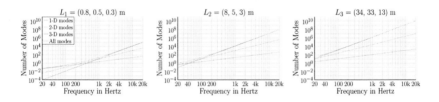

Figure 4.14.: Number of 1-D, 2-D and 3-D modes and total number of modes over frequency for three different exemplary rooms with geometries L_1, L_2 and L_3.

For the high frequency range, where the computational complexity becomes impractical due to the number of modes, the image source method could be employed instead. This method yields identical results to the modal superposition for rectangular rooms without damping as proven in (ALLEN and BERKLEY, 1979). But this requires theoretically an infinite number of image sources up to infinite image source order. The limitation of this order, which is practically unavoidable, yields errors especially at the lowest frequencies where the modal behavior is dominating. Details on these effects are presented in (ARETZ, DIETRICH, and VORLÄNDER, (2013)). Furthermore, the combination of results obtained from both approaches can be found in (ARETZ, 2012). Only the modal superposition is used in the following.

The number of 3-D modes exceeds the number of 2-D and 1-D modes for increasing frequencies as shown in Figure 4.14 for three different exemplary rooms with $L_1 = (0.8\,\text{m}, 0.5\,\text{m}, 0.3\,\text{m})$, $L_2 = (8\,\text{m}, 5\,\text{m}, 3\,\text{m})$ and $L_3 = (34\,\text{m}, 33\,\text{m}, 1.23\,\text{m})$[9]. The Propability Density Function (PDF) of $\psi_i(\mathbf{r})$ for all positions in the room (uniform PDF is used for $\mathbf{r}$) is different for these three classes of modes as depicted in Figure 4.15. For a fixed source position the modal coefficients for 1-D modes are likely to be close to their extreme values ± 1 for arbitrary receiver positions inside the room[10]. For this fixed source the position in a corner has to be considered as a worst scenario as the modal coefficients ψ_i for the source are all at its extreme values. For the mixed modes (2-D, 3-D) it is more likely that the coefficient of a single modes is close to zero and hence, the mode has no impact on the final result. For arbitrary source and receiver positions the PDF narrows around zero.

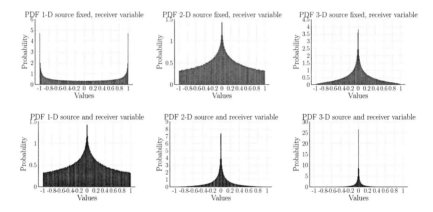

Figure 4.15.: Probability of the values of the modal coefficients ψ_i for 1-D, 2-D and 3-D modes for fixed (top) and variable (bottom) source position and variable receiver position based on normalized histogram from Monte Carlo Simulation.

For the fitting of impulse responses by the model described in Section 2.3.3 this behavior can be exploited to reduce the number of modes (poles) to be fitted. Furthermore, this behavior can also be used to decrease the computation time as only a smaller number of modes has to be used for the modal superposition. Neglecting poles with low energy contribution will introduce a small error. This error can be estimated by introducing a threshold c_{limit} and considering modes

---

[9]The rooms with L_1, L_2 and L_3 correspond to the ITA *auralization box* (DIETRICH et al., 2010), the reverberation chamber of the ITA, and the Eurogress, multi-purpose hall in Aachen, respectively.

[10]The terms *source* and *receiver* can also be exchanged in this context.

with higher absolute modal coefficients only. The relative error can be estimated by accounting for the energy of the neglected modes in relation to the total energy (HENSE, 2012). This behavior is given in Figure 4.16 for a distribution of the modal coefficients for 3-D modes and variable source and receiver position. As an example, the number of modes can be reduced to approx. 20 % and hence the computation time can theoretically be also reduced to 20 % by allowing a relative error of 5 %.

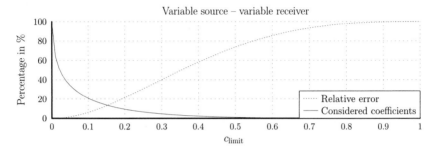

Figure 4.16.: Relative energetic error due to neglecting of modes with modal coefficients smaller than the absolute threshold value c_{limit}.

As the eigenfrequencies are not dependent on the position of the source and the receiver in the room, the applicability of the Common Acoustic Poles and Zeros approach (CAPZ) is given as the resonances or the poles are common for all positions and the residues and hence also the zeros—here in particular the residues in terms of modal coefficients—are dependent on the position.

This analytic model has been extended towards arbitrary directivity patterns for the source and the receiver in (POLLOW, DIETRICH, and VORLÄNDER, 2013). For physical multipoles of higher order than the monopole the eigenfunctions are substituted by derivations. The multipoles are described by (k,l,m)-th derivation in x, y, z-direction, respectively as

$$\psi_{i,k,l,m}(\mathbf{r}) = \frac{\partial^k}{\partial x^k} \frac{\partial^l}{\partial y^l} \frac{\partial^m}{\partial z^m} \cdot \psi_i(\mathbf{r}). \qquad (4.18)$$

E.g., the dipole in x-direction reads as

$$\psi_{i,1,0,0}(\mathbf{r}) = \frac{\partial}{\partial x} \cdot \psi_i(\mathbf{r}). \qquad (4.19)$$

This model is validated by a comparison with numerical simulation using the Boundary Element Simulation (BEM) in *LMS Virtual.Lab 11 SL-2* up to a frequency of 1000 Hz with a resolution of 1 Hz. The maximum distance between two nodes was 20 mm. Hence, the theoretical frequency limit of the simulation was 2.8 kHz and safely above the maximum frequency simulated. The room dimension are 80 cm × 50 cm × 30 cm, with the source placed at $\mathbf{r_s} = (15\,\text{cm}, 15\,\text{cm}, 15\,\text{cm})$ and the receiver at $\mathbf{r_r} = (0, 0, 0)$. No absorption was applied in the BEM and no modal damping was applied for the analytic model. A monopole receiver was used with a monopole and a dipole (x-direction) source. The analytic calculation according to Eq. 4.13 was performed up to a frequency of 10 kHz (13490 eigenfrequencies, $\Delta f = 1\,\text{Hz}$, $f_s = 44100\,\text{Hz}$)[11]. The comparison of the room transfer function obtained with both simulation methods is shown in Figure 4.17 where the results show very good agreement in both magnitude and phase up to the maximum frequency in the BEM.

For higher orders than the dipole the frequency limitation and hence the finite number of modes used causes a problem. The modulus of the modal coefficients rises over frequency. The Frequency Response Function (FRF) of a single mode at higher frequencies causes an offset at low frequencies. In case the modal superposition was conducted up to infinite frequencies, this offset vanishes as it cancels out with the offset caused by other modes. The implementation used in the following solves this stability problem by using two simulations. The first FRF is obtained up to the desired frequency limit and the second FRF includes only the first few modes. The residue at the frequency $f = 0\,\text{Hz}$ is then matched.

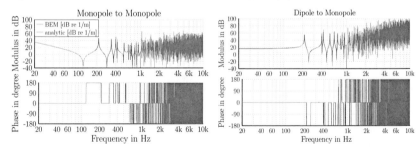

Figure 4.17.: Comparison of room transfer function calculated by the BEM and the analytic model for an undamped rectangular room with dimensions 80 cm × 50 cm × 30 cm for monopole (left) and dipole (right) sound source and monopole receiver.

The spatial derivatives of the monopole result in the *physical multipoles* (WILLIAMS, 1999). These physical multipoles are, e.g., the dipole and quadrupole but they do

---

[11] Computation time approx. 5 s on *Apple Mac Book Pro*, 2 GHz, Core i7, 8 GB RAM, MATLAB R2012b, *Mac OS X Mountain Lion*

not directly correspond to the SH base functions for orders $n \geq 2$ as used in the previous sections. Figure 4.18 shows the monopole with $(k, l, m) = (0, 0, 0)$, dipole $(1, 0, 0)$, the longitudinal $(2, 0, 0)$ and the lateral $(1, 1, 0)$ and $(0, 1, 1)$ quadrupoles. There exist 6 different quadrupoles in the physical multipole representation, where only these two fundamental shapes occur. The SH representation contains only 5 coefficients in the same order. The shapes of the physical multipoles do not directly match the shapes of the complex SH base functions shown in Figure 4.1. A set of transfer functions for one combination of source and receiver position for a receiver represented by physical multipoles up to an order n_{max} to a monopole receiver can be transformed to a set of transfer functions for the receiver represented by SH base functions up to the same order. This transformation is not unique as there exist more multipole coefficients in one order than spherical harmonics coefficients (WILLIAMS, 1999). The eigenfrequencies for sources and receivers of higher orders remain the same, but the coefficients weighting the contribution of the modes at a particular position changes.

As loudspeakers, e.g., the dodecahedron measurement loudspeaker commonly used in room acoustics, show deviations from the omni-directional pattern for higher frequencies, i.e., $f \geq 1\,\mathrm{kHz}$, this model is not suitable for the prediction of measurement uncertainties in room acoustics due the large number of modes and hence long computation times for typical room dimensions larger than the dimension in the previously used example of the small chamber. As mentioned earlier, the Image Source Method (ISM) could be used instead as, e.g., in the ongoing work of WITEW (KNÜTTEL, I. B. WITEW, and VORLÄNDER, (2013); I. WITEW et al., 2013; I. WITEW, KNÜTTEL, and VORLÄNDER, 2013) for the upper frequency range also including the directivity pattern.

(a) $(0,0,0)$	(b) $(1,0,0)$	(c) $(2,0,0)$	(d) $(1,1,0)$	(e) $(0,1,1)$

Figure 4.18.: Physical multipoles up to the order $n = 2$: (a) Monopole, (b) dipole, (c) longitudinal quadrupole and (d) horizontal and (e) vertical lateral quadrupole.

4.3.2. Remark on the Derivation of Room Acoustic Parameters

Although the formulation of the analytic model is in frequency domain the room impulse responses can be obtained and hence room acoustic parameters according to ISO 3382 can be calculated. The Reverberation Time (RT) of each mode can be controlled by the modal damping constant. The RT observed in a room can be considered as a mix of the modal reverberation times. At first, it is investigated how the reverberation time according to ISO 3382 relate to the modal reverberation times in case they are all set to the same value of 1 s.

In order to provide more information the signal number parameters are obtained for different frequency bands. In this context octave or third-octave bands are commonly uses. The band limitation is applied to the impulse response prior to the evaluation of the parameter for each frequency band separately. As the standard does not provide a particular algorithm or even implementation for this filtering, the results obtained by different software although compliant to the ISO standard can vary significantly. Moreover, the band filters can introduce a systematic error to the obtained reverberation time that becomes problematic for short reverberation times of the room (KOB and VORLANDER, 2000). Third-octave bands are mainly used in the following with an implementation of fractional octave band filters compliant to the ISO standard. The uncertainties introduced by the band filters are beyond the scope of this thesis.

The reverberation times EDT, T_{10}, T_{20} and T_{30} using different lengths of the decay curve for the linear regression algorithm are shown in Figure 4.19. All reverberation times obtained by the algorithm according to the standard show deviations from the ideal value of 1 s over the entire frequency range. For increasing length of the portion of the decay curve used (EDT to T_{30}) the deviations decrease (DIETRICH et al., 2013). Moreover the deviation decreases for increasing numbers of modes within the frequency band for evaluation and hence the deviation decreases with increasing frequencies due to the rising number of modes per band. This deviation is neither caused by errors in the analytic model nor by errors in the implementation of the algorithm for the calculation of the reverberation times. The specified algorithm in the standard itself causes this deviation. However, this behavior is the same for the evaluation of real measurement data. It is not in the scope of this thesis to develop a new formulation of the algorithm. Although it is commonly assumed that the reverberation time in the diffuse field is independent on the position (KUTTRUFF, 2000), the reverberation times obtained by this algorithm will vary with changes in the position. Details on the roots of this problem are presented in the appendix A.3.

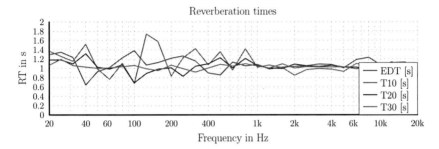

Figure 4.19.: Reverberation times obtained according to ISO 3382 for a room transfer function with ideal modal reverberation times of 1 s. Monopole source and receiver at $r_s = (1.5\,m, 1.5\,m, 1.5\,m)$ and $r_r = (8\,m, 5\,m, 3\,m)$ and modal superposition up to a frequency of 9 kHz.

4.3.3. Parametric Model and Rational Fitting

As the rectangular room model is a basic model, the relation to more complex room geometries and especially the relation to measurement data has to be investigated. The application a pole-zero fitting method to room acoustics has been presented in (MOURJOPOULOS and PARASKEVAS, 1991). An exemplary measured room impulse response from a measurement session at the Eurogress hall in Aachen[12] is fitted by the vector fitting as described in Section 2.3.3. This hall is not rectangular, has several scattering objects and a balcony.

A set of modal parameters is obtained containing the eigenfrequency f_i, the damping constant δ_i or its associated reverberation time of the mode, and the modal coefficient c_i. Only the combination c_i of both coefficients ψ_i for the source and the receiver can be obtained by a fit. Only if the exact position of the source and the receiver and the exact shape of each mode is known, these parameters could be theoretically separated. Usually, room shapes deviate strongly from the rectangular room and the modal shapes are generally not known. Hence, this separation is impractical and only the combination is shown in the following.

---

[12]Measurements conducted by WITEW, DIETRICH, KNAUBER and first results published in (I. WITEW and VORLÄNDER, 2011))

By using Eq. 2.11 and the approximation in Eq. 4.17 the maximum frequency that could be reasonably used for the fit can be estimated as

$$f_{max} \approx \sqrt{\frac{3T_{max}c^3}{4V\pi}} \qquad (4.20)$$

where T_{max} is the length of the useful part of impulse response. For measurement data this length can be approximated by the reverberation time RT and the peak SNR as

$$T_{max} = \frac{SNR \cdot RT}{60\,dB} \; . \qquad (4.21)$$

This is equivalent to the intersection time of the impulse response with the noise floor. The length of the impulse response could be virtually extended by zero-padding. However, this method does not improve the quality of the data but is an interpolation method only and generates additional frequency bins with correlation to the surrounding bins.

The volume of the Eurogress hall is approximated as $V = 34\,m \times 33\,m \times 13\,m$ with three lengths based on an equivalent rectangular room. The number of modes up to an arbitrary maximum frequency was shown in Figure 4.14. According to Eq. 4.20 the maximum frequency possible for the fit is determined as $f_{max} = 34\,Hz$ and the number of considered poles is 62. On the first sight, this allows the investigation of the very low audible frequency range only. By taking into account the PDF of the modal coefficients as illustrated in Figure 4.15, the relative error according to Figure 4.16 and applying zero-padding to virtually extend the resolution to $T_{max} = 6\,s$ the frequency range can be extended (HENSE, 2012).

The extended frequency is chosen as $f_{max} = 500\,Hz$ and only a fraction of the theoretical number of modes is used. The percentages used are 0.1 % (196 modes), 0.5 % (980 modes) and 1 % (1960 modes), which is close to the theoretical number of modes for this maximum frequency in this particular room[13]. By comparison with Figure 4.16, large energetic errors are expected for at least the fit using only 0.1 % of the theoretic room modes. The modal parameters are inserted into the parametric model for ongoing modal superposition instead of the parameters previously obtained from the rectangular room. The magnitude spectrum of the measurement and fitted model along with the deviation is shown in Figure 4.20. Up to the maximum theoretic frequency the fit agrees with the shape of the original spectrum. Above this frequency the FRF of the fit tends to zero with increasing frequency. The spectral division by the measured data shows the

---

[13]The computation time on *Apple Mac Book Pro*, 2 GHz, Core i7, 8 GB RAM, MATLAB R2012b, *Mac OS X Mountain Lion* approx. 1 min, 10 min and 100 min, respectively

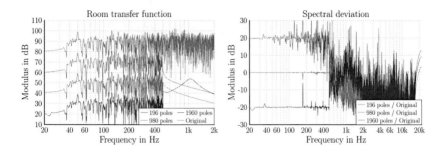

Figure 4.20.: Comparison of measured room transfer function and rational fit method with different numbers of poles up to frequency of 500 Hz (left) and spectral deviation (right) in Eurogress hall, Aachen. Results are shifted by 20 dB.)

deviation in detail. For the fit with 196 modes the deviations are around ±5 dB up to a frequency of approx. 100 Hz. This deviation increases over the frequency range of interest up to a value of approx. ±20 dB as the modal density increases as well. The number of modes is to low to capture the dominating modes in the frequency range. The resulting FRF with 960 modes shows small deviations in the range of ±1 dB over the entire frequency range of interest except for a narrow band deviation. The result using 1960 modes shows larger deviations than the previous one although the accuracy is expected to be higher especially according to the results in Figure 4.20. This can be explained by a decreased convergence of the fitting algorithm, as the degree of freedom for the algorithm rises and finally becomes problematic. This could be compensated by using a sub-band fitting approach reducing the number of modes for each fit and finally combining the fitted data from these frequency bands.

The room acoustic parameters EDT and C_{80} calculated from the original and the fitted impulse response are shown in Figure 4.21. These results show good agreement regarding the order of magnitude for all number of modes investigated. As already indicated by the spectral deviation the closest match is achieved with 980 modes. But even for only 0.1 % of the modes considered the room acoustic parameter are still roughly captured. As the maximum frequency falls on the center frequency of the highest third-octave band, this band is not fully covered by the fit and hence the room acoustic parameters deviate. Although, results are only shown for one position and one room, the applicability of the fitting algorithm to room acoustic measurement data and hence to also the relation to modal superposition approach is hereby shown.

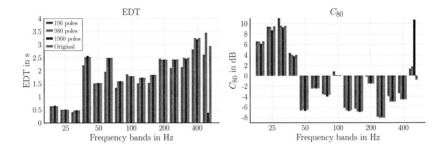

Figure 4.21.: Room acoustic parameters EDT and C_{80} of the measured impulse response compared to the fitted responses with different numbers of modes.

4.4. Application II—Microphone Position and Source Orientation

The analytic model is used to investigate the influence of the position of the monopole receiver for a monopole source and the orientation of basic radiation patterns for the source and a fixed monopole receiver. Moreover, the characteristics of the modal coefficients are investigated and the evaluated room acoustic parameters are discussed.

4.4.1. Microphone Position and Room Acoustic Parameter

The same grid as later used in the measurements is chosen as $2.35\,\mathrm{m} \times 2.05\,\mathrm{m}$ with a 5 cm distance between the microphones for the room geometry L_3. The source is placed in the corner to include all modes and the grid is positioned in the center of the room with a height of $1.2\,\mathrm{m}$. The maximum frequency is $f_{\mathrm{max}} = 562\,\mathrm{Hz}$. Hence, the room acoustic parameter up to the 500 Hz third-octave band can be evaluated. To obtain realistic results the measured mean reverberation time of the Eurogress hall is used for each third octave band individually.

Figure 4.22 shows the evaluated room acoustic parameters EDT and C_{80} for the third-octave bands 125 Hz, 250 Hz and 500 Hz. The parameters are dependent on the position and show a variance over the area that increases with increasing frequency. These findings have already been shown in (DIETRICH, 2007) for a smaller array of $40\,\mathrm{cm} \times 40\,\mathrm{cm}$ based on measurement data only. This sim-

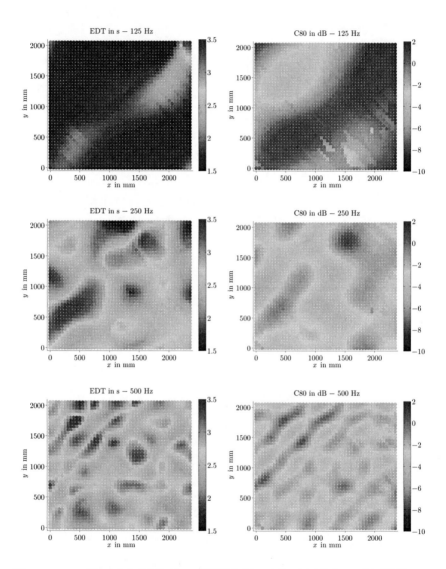

Figure 4.22.: Simulated distribution of EDT (left) and C_{80} (right) for a grid of $2.35\,\mathrm{m} \times$ $2.05\,\mathrm{m}$ for the third octave bands $125\,\mathrm{Hz}$, $250\,\mathrm{Hz}$ and $500\,\mathrm{Hz}$ using an analytic model for rectangular rooms.

ulation model does not include noise or other artifacts as usually observed in measurements. Hence, the variation observed is clearly proven to be caused by the variation of the position. The same model has been used to investigated the standard deviation over the area for different sizes of rooms in (GUSKI, DIETRICH, and VORLÄNDER, 2012). The statistical analysis of the variance over space in reverberation rooms has already started in the early 80s by (DAVY, 1981).

Figure 4.23.: Room acoustic measurement at Eurogress hall, Aachen with 4 three-way dodecahedron loudspeakers on the musical stage (left) and microphone scanning setup with 24 *Sennheiser KE4* microphones.

4.4.2. Comparison with Room Acoustic Measurements

An automated measurement of room impulse responses with a step motor controlled linear microphone array with 24 Sennheiser KE4 electret condenser microphones of a an area of 2.35 m × 2.05 m in the Eurogress multi-purpose hall in Aachen was conducted (I. WITEW et al., 2010). The measurement was done sequentially resulting in a microphone grid with minimum distances of 5 cm. The height of the microphone positions was approx. 1.2 m high above the floor as defined in ISO 3382. Only results for the second dodecahedron loudspeaker from the left shown in Figure 4.23 are shown in the following.

The mean and the standard deviation of both room acoustic parameters are depicted in Figure 4.24. For the reverberation time the relative standard deviation is used. For the low third-octave bands the standard deviation is higher than for the high frequency bands.

The shape of hall is not rectangular. Figure 4.26 shows the distribution in space for the low third octave bands around 125 Hz, 250 Hz and 500 Hz. The

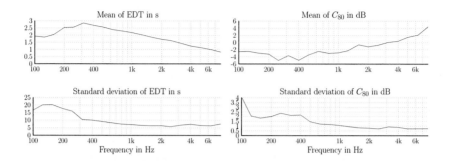

Figure 4.24.: Mean of the room acoustic parameters EDT and C_{80} over the scanned area in the Eurogress hall, Aachen and the corresponding (relative) standard deviation over frequency.

fluctuation of the parameters over the space increase with increasing frequency and is hence an effect which is clearly related to the extent of the wavelength. The measurement was conducted in 3 steps with a manual shift of 70 cm in y-direction. As can be seen in the plots, this shift was not as accurate as the movements by the step-motor. Hence, a mismatch of approx. 5 cm can be assumed based on the discontinuities observed along the y-direction at 70 cm and 140 cm.

The principal shape of the distribution over the area and the oscillation of the parameters in the different frequency bands are similar to the simulation results, although the simulation used a basic model with a rectangular room. However, the parameters do not match in detail but this is not problematic as only the deviations and hence the uncertainties should match.

Figure 4.25 shows the measured deviation of both parameters up to 8 kHz in selected third-octave bands in the Eurogress hall with geometry L_3. This plot comprises a comparison of the room acoustic parameters obtained at all positions within the array. Hence, the maximum distance is approx. 3 m. For the absolute distance d in meters the increasing inclination of the curves with frequency is observable. The normalized distance as the product of the absolute distance with the wavenumber kd compensates this behavior as shown. With this notation the distance of exactly one wavelength is represented by the value 2π in the normalized distance. In particular, two disturbing effects are additionally observed in the relative plots. Firstly, the limited size of the scan grid results in a missing asymptotic behavior for the 63 Hz and 125 Hz curve. Secondly, the limited resolution of the grid in terms of minimum microphone distance, limits

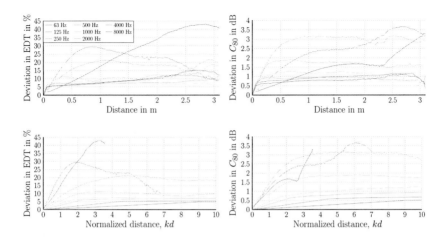

Figure 4.25.: Measured standard deviation of EDT (left) and C_{80} (right) over the absolute distance in meters (top) and the normalized distance (bottom) in selected third-octave bands. These plots capture the uncertainty of the room acoustic parameters over the uncertainty in positioning.

the resolution of the curve for short wavelengths and therefore the 4 kHz and 8 kHz frequency band. Theoretically, the microphone distance is almost sufficient to capture one kd for the upper frequency band but the evaluation algorithm further requires additional values for averaging the reasonable resolution for the plots is even lower. Hence, the interesting part indicating the inclination of the curve and the point when a constant deviation is reached is not observable for these frequency bands.

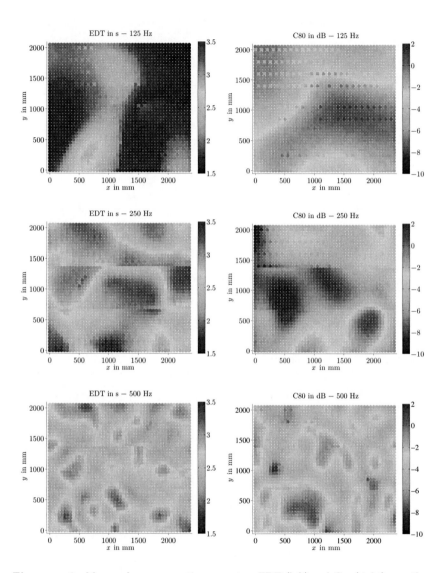

Figure 4.26.: Measured room acoustic parameters EDT (left) and C_{80} (right) over the scanned area of $2.35\,\text{m} \times 2.05\,\text{m}$ in Eurogress hall, Aachen for $125\,\text{Hz}$, $250\,\text{Hz}$ and $500\,\text{Hz}$ third-octave bands.

4.4.3. Uncertainty Modeling—Position of Receiver

As the observed deviations of the room acoustic parameters are dependent on the wavelength as shown in the previous result, the link with the variation of the modal coefficients is investigated. Changes in the temperature cause changes of the speed of sound. This influences the eigenfrequencies only if the relative position in the room is kept constant. Hence, the modal parameters remain the same. If the room boundaries are furthermore assumed to have frequency independent characteristic, the variation of the temperature and the variation of the position in the room can be investigated separately. Changes in temperature are investigated in the appendix A.4.

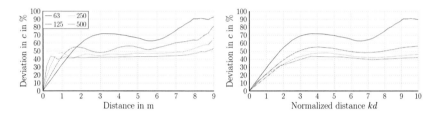

Figure 4.27.: Standard deviation of the modal coefficient over the distance (left) and the normalized distance as the product of absolute distance and wavenumber (right) averaged over all modes in the third octave band and all positions in the room L_2. These plots capture the uncertainty in the modal coefficient if the distance is considered as the uncertainty of the position.

The fluctuation of the modal parameter over the space is strongly dependent on the mode number and hence also on the associated eigenfrequency. In case of the rectangular room model and the cosine terms this fluctuation can be considered in terms of a periodicity with the wave number $k_i =$ of this modes. This is not obvious on the first sight as the modal coefficient is not directly linked with the wave number or the eigenfrequency in Eq. 4.14 but only the mode number n. The three ratios of the mode numbers and the room geometry for each Cartesian coordinate are summarized for the calculation of the eigenfrequency. Moreover, the three cosine terms could also be summarized in a similar manner yielding a dependence that is proportional to the wavelength.

A scan of the room is conducted in terms of MC simulations using 500 random receiver positions with a uniform distribution over the entire volume of the room. It is important to mention that the scan area or volume has an impact on the results as shown in the appendix A.5.

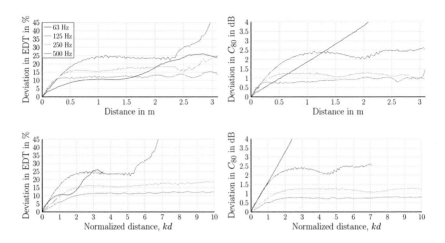

Figure 4.28.: Simulated standard deviation of the room acoustic parameters over the distance in Eurogress hall with geometry L_3 for the scan grid geometry as used in the measurements.

Figure 4.27 shows the deviation of the modal coefficient over the distance. This deviation can be understood as an uncertainty in the modal coefficient (y-axis) for a given uncertainty in the position (x-axis). Furthermore, it describes in how far a measurement at one position is representative for a finite area around this position. In the low frequency bands the deviation rises slowly over the distance. For the high frequency bands, the deviation rises faster and reaches an asymptotic plateau. The normalized distance kd gives more insight into this behavior. The plateau is reached at values of 3-4 for kd and is around $1/2$ of a wavelength. Morever, the value of this plateau decreases with increasing frequency. This is caused by the increasing number of modes in a frequency band. Hence, more and more modes cancel out this oscillating behavior. The periodicity of the oscillation can be estimated to 2π over the normalized distance from the plots. The behavior is investigated in the appendix A.6.

All curves show a tail that is rising towards the maximum distance observable in the room. This effect is caused by the reduced number of points available that have this maximum distance. For the room L_2 as much as 6, 28, 168 modes fall with their eigenfrequencies in the 63 Hz, 125 Hz and 250 Hz third octave band, respectively. It can be concluded that the modal coefficients reach a maximum deviation of approx. $40 - 60\%$ for values of kd around 3-4 corresponding again to approx $1/2$ of a wavelength. Hence, this is a measure indicating that the transfer functions can also be expected to show a maximum deviation accordingly.

These findings are in agreement with the measurements in the Eurogress hall, especially the deviation of the room acoustic parameters over the distance in Figure 4.25 where the deviation of the simulated room acoustic parameters is depicted in Figure 4.28. At approx $1/2$ of a wavelength a constant deviation of the room acoustic parameters is found. Moreover, the room acoustic parameters seem to settle even at bit earlier at a value of 3 for kd than the modal coefficients. Hence, a link between the uncertainties in the modal coefficient, the room transfer function and also the derived room acoustic parameter is found. These results are in good agreement with the findings in (PIERSOL, 1978). He shows that the coherence of two room transfer functions with a distance d are proportional to $\sin(kd)/kd$. Hence, the coherence drops to values close to zero for kd between 0.5 and 1.

It is important to mention, that the absolute uncertainty in position is normalized to the room geometries in Eq. 4.14. Hence, the uncertainty of the modal coefficient decreases with increasing room sizes. By looking at a fixed frequency, this effect is overcome by the increasing mode numbers for increasing room sizes.

It is beyond the scope of this thesis to investigate the relevance of the observed deviations in psycho-acoustical terms and it is referred to a short study involving listening test. The stimuli were obtained using this model with the approach introduced above (MASIERO et al., 2012).

4.4.4. Uncertainty Modeling—Orientation of Source

The enhanced model from Section 4.3.1 is used to investigate the effect of the rotation of ideal dipole and quadrupole sources on the room impulse response and the room acoustic parameters.

A set of room impulse response for the monopole, the dipoles and the quadrupoles in the physical multipole representation have been calculated. The chosen room geometry is $L_2 = (8, 5, 3)$ m. The source and receiver position are chosen as $r_s = (1.5\,\text{m}, 1.5\,\text{m}, 1.5\,\text{m})$ and $r_r = (8\,\text{m}, 5\,\text{m}, 3\,\text{m})$ and the maximum frequency as $f_{\text{max}} = 9\,\text{kHz}$[14]. The transfer function is depicted in Figure 4.29. The monopole has a fairly flat energy distribution over frequency where as the dipole and the quadrupole spectra show more and more increasing levels over frequency due to the differentiation in Eq. 4.14. E.g., a differentiation in x yields a factor of $\pi n_{x,i}/L_x$ which increases over frequency as the modal number increases over frequency.

The sources with the basic patterns are rotated around their vertical axis. All radiation patterns that are rotationally symmetric around the z-axis will yield results that are independent on this rotation. Hence, the dipole $(0, 0, 1)$ and the longitudinal quadrupole $(0, 0, 2)$ are not considered. Details on the rotation of the dipole and quadrupole without using SH and the Wigner-D matrix is provided in the appendix A.2. As can be seen from these equations the relative deviation in the modal coefficients is independent on frequency. E.g., the modal coefficient for a mode around 100 Hz deviates in the same manner as the modal coefficient for 10 kHz for the same rotation angle. This seems to be in contradiction with the observed results in room acoustics in the past. But this behavior can be explained as in real-life measurements the radiation pattern and hence the influence of dipole and quadrupole patterns—but especially also higher orders that are not considered in the following—becomes stronger with increasing frequency. In the simulation results shown these basic patterns are investigated separately and hence the dependence over frequency is not captured. In order to obtain similar results as observed in the measurements a frequency weighted superposition of the rotation of these patterns (monopole, dipole, quadrupole, etc.) had to be conducted.

---

[14]Computation time approx. 2.5 h on *Apple Mac Book Pro*, 2 GHz, Core i7, 8 GB RAM, MATLAB R2012b, *Mac OS X Mountain Lion*.

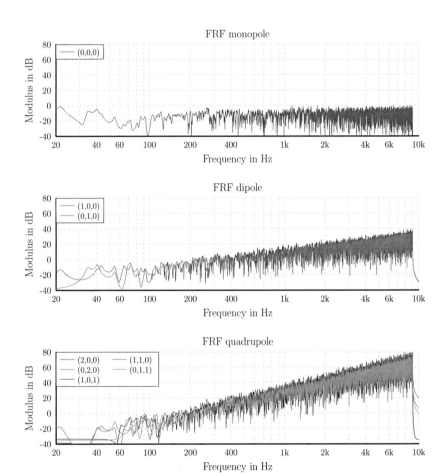

Figure 4.29.: Room transfer function for L_2 with monopole, dipole and quadrupole source radiation pattern. The source and receiver position are $\mathbf{r_s} = (1.5\,\mathrm{m}, 1.5\,\mathrm{m}, 1.5\,\mathrm{m})$ and $\mathbf{r_r} = (8\,\mathrm{m}, 5\,\mathrm{m}, 3\,\mathrm{m})$ with RT = 1 s.

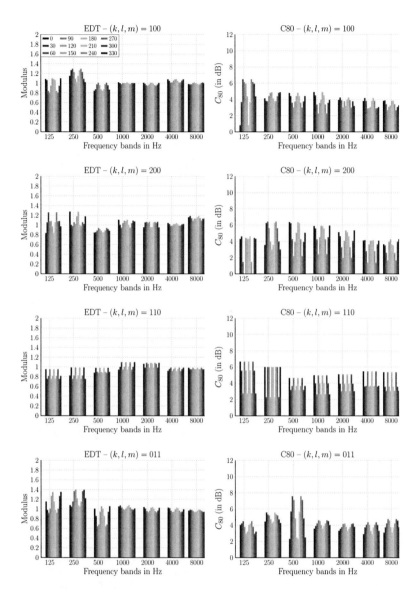

Figure 4.30.: Room acoustic parameters EDT and C_{80} for 360° rotation of dipole and quadrupoles denoted by the vector (k, l, m) in 30° steps (color indicating the rotation angle).

The relative deviation of the modal coefficients for the basic multipole patters due to a rotation of the source is hence only dependent on the specific pattern and the rotation angle.

Figure 4.30 shows the room acoustic parameters obtained from a $360°$ rotation in $30°$ steps of the multipoles discussed. The rotational periodicities can be observed as $180°$ for the dipole and the longitudinal $(2, 0, 0)$ and the lateral $(0, 1, 1)$ quadrupole. The lateral $(1, 1, 0)$ quadrupole shows a periodicity of $90°$ which corresponds to the equations used for the rotation. The rotation of the dipole by $180°$ yield an inverted frequency response, but the relation of all modal coefficients is already the same for $180°$. The evaluation of the room acoustic parameters also yields the same results already for $180°$ instead of $360°$ since the room acoustic parameters are energy-based and use the squared impulse response. Hence, the sign of the impulse response does not have any influence on the result. A rotation of the lateral $(1, 1, 0)$ quadrupole already yields the inverted frequency response at an angle of $90°$ and hence, the room acoustic parameters are already the same for this angle. In contrary the periodicity of the radiation pattern of the longitudinal $(2, 0, 0)$ quadrupole is $180°$ and equivalent to the periodicity observed in the room acoustic parameter since there is no inverted response due to only positive radiation. The periodicity in the room acoustic parameters can be smaller than the periodicity of the multipole pattern over the rotation angle but this is not necessarily the case as shown.

As can be seen in the room acoustic parameters in Figure 4.30 the variation over the rotation angle is also dependent on frequency although each modal coefficients c_i is not. This can be explained by the increasing modal density over frequency and also the different interaction between the room modes in the different frequency bands. The relative deviation of the mean of all all modal coefficients in a frequency band for the given room over the uncertainty of the orientation angle is shown in Figure 4.31. As can be seen, even the mean including the influence of the modal density is only slightly dependent on the frequency.

MC simulations were run to investigate the uncertainty in the room acoustic parameters in the different frequency bands if the source orientation in terms of its orientation around the z-axis is uncertain. The simulation uses 20 runs with arbitrary source orientations of the physical multipoles. This orientation is used as a reference for the run. Additional simulations are then run with different uncertainties of this exact orientation angle. The room acoustic parameters are evaluated and the mean of the absolute error for each difference in angle over all iterations is calculated. This results in the uncertainty of the room acoustic

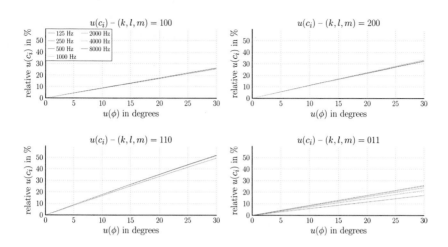

Figure 4.31.: Relative deviation in modal coefficient for different frequency bands and dipole and quadrupole pattern over uncertainty of rotation angle $u(\phi)$.

parameter for different uncertainties in the orientation. In terms of the GUM procedure this is the sensitivity. The results are presented in Figure 4.32 where the three lowest octave bands 125 Hz, 250 Hz and 500 Hz show mostly the highest deviations. In these frequency bands the modal density is low compared to the high frequency bands. I.e., although the influence of the higher multipole orders increases over frequency and is hence contrary to the results shown, the combination of both as observed in real-life measurements still yields increasing uncertainties of the room acoustic parameters with increasing frequency.

In order to stay below, e.g., the Just Noticeable Difference (JND) according to ISO 3382 for the clarity index of 1 dB a maximum uncertainty of $u(\phi) = 15°$ is allowed for both dipole and quadrupole. The JND of 5 % for the reverberation time with a mean RT of 1 s is reached with $u(\phi) = 26°$ for the dipole and $u(\phi) = 8°$ for the quadrupole. A similar settling behavior as found for the position uncertainty can not be stated. However, the combination of these spherical base patterns for higher orders and hence synthesizing more realistic results comparable to measurements with directive loudspeakers should be investigated. Depending on the combination of the base functions such a settling behavior might be found.

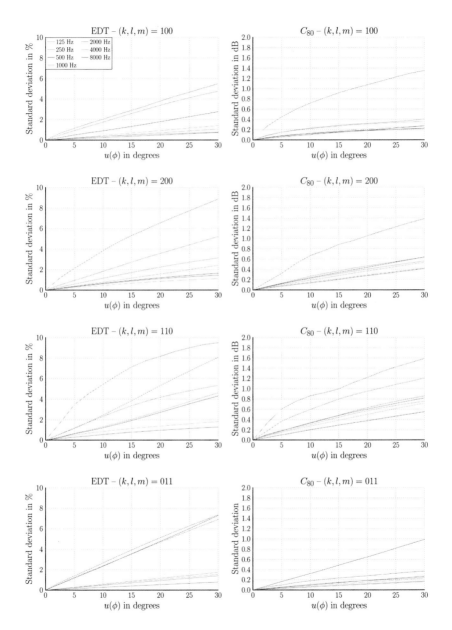

Figure 4.32.: Standard deviation of room acoustic parameters EDT (left) and C$_{80}$ (right) for mismatch in orientation angle around the z-axis for different basic physical multipoles.

4.5. Measurement of Transfer Functions for Variable Source Directivities

In case the acoustic scenario cannot be modeled by an analytic expression, either because the computational complexity is to high or the scenario cannot be approximated with sufficient accuracy, the need for a measurement method rises. This should still preserves the advantageous of an parametric model in order to allow uncertain input parameters in a post-processing step. A set of impulse responses with different radiation patterns of the sound source (or the sound sensor) can be measured. This approach is still considered as parametric in a way, as the source (or sensor) can be positioned virtually after the measurement has been completed within a certain limits. In particular, the measurement loudspeaker might be substituted virtually by a different loudspeaker with different radiation characteristics and this loudspeaker might even be placed at a different position but in the vicinity of the original measurement position in the room. The link between radiation characteristic and translational shift is presented in (ZOTTER, 2009) but not further considered in this thesis.

4.5.1. Developed Measurement Method

The method developed in this context is motivated by (ZOTTER, 2009)[15]. But instead of measuring a room impulse response with a loudspeaker array that emulates the particular radiation characteristic of interest during the measurement, the approach is more general as the synthesis of the specific radiation pattern is done after the measurement. However, the mathematical background required is identical. The method itself has been firstly implemented by (KUNKEMÖLLER, 2011) and summarized in (KUNKEMÖLLER, DIETRICH, and POLLOW, 2011; POLLOW et al., 2011).

A mid-range dodecahedron loudspeaker with a radius of approx. $15\,\mathrm{cm}$ with $L = 12$ independently driven membranes has been used as spherical loudspeaker array. However, the method is applicable to arbitrary loudspeaker geometries and arbitrary number of membranes. The directivity pattern of each membrane l has to be measured in the acoustic free-field expressed in its SH coefficients in the far-field for each frequency summarized in a vector $\hat{\mathbf{d}}_l$. This vector is

---

[15]Parts of the following section have been published in (POLLOW et al., 2012)

organized stepping through all orders up to a SH harmonic order n and for each order through all available degrees m as

$$\hat{\mathbf{d}}_l = \left(\hat{d}_{0,l}^0, \hat{d}_{1,l}^{-1}, \ldots \hat{d}_{1,l}^1, \ldots, \hat{d}_{n,l}^{-m} \ldots, \hat{d}_{n,l}^m \right)^T . \tag{4.22}$$

The directivities of all membranes are summarized in the directivity matrix $\hat{\mathbf{D}}$ in SH as

$$\hat{\mathbf{D}} = \left(\hat{\mathbf{d}}_1, \hat{\mathbf{d}}_2, \ldots, \hat{\mathbf{d}}_l \ldots, \hat{\mathbf{d}}_L \right) . \tag{4.23}$$

Each room acoustic measurement with this array yields L impulse responses h_l with corresponding directivity $\hat{\mathbf{d}}_l$. In order to increase the maximum harmonic order, the loudspeaker array can be rotated to different orientation angles and further information can be gained in terms of impulse responses $h_{l,i}$. The corresponding directivities $\hat{\mathbf{d}}_{l,i}$ can be obtained by using the Wigner-D rotation matrix as mentioned earlier.

The method superposes the measured impulse responses to obtain a room impulse response h_T for a target directivity pattern $\hat{\mathbf{p}}_T$. A numerical solutions can be obtained by using the Moore-Penrose pseudo-inverse along with Tikhonov regularization specified by the regularization parameter $\varepsilon > 0$ as

$$\mathbf{w}_T = \left(\hat{\mathbf{D}}^H \hat{\mathbf{D}} + \varepsilon \mathbf{I} \right)^{-1} \hat{\mathbf{D}}^H \hat{\mathbf{p}}_T \tag{4.24}$$

where $\mathbf{I}$ is identity matrix. This weighting vector $\mathbf{w}_T$ is applied to each frequency bin of the measured impulse responses and followed by superposition to obtain the room impulse response h_T. Especially for room acoustic analysis it is important to mention that $\hat{\mathbf{p}}_T$ can be chosen in a way to obtain room impulse responses for SH base functions, e.g., a room impulse response for the monopole and hence the omni-directional radiation pattern. This can is achieved with a sound source with distinct directivity rather than omni-directional directivity.

4.5.2. Limitation of the Method

The method has two limitations—one regarding the maximum frequency and one regarding the maximum SH order. Both are again linked by Eq. 4.7. In order to analyze the theoretical limitations and the applicability the summarized error in the SH orders over frequency is depicted in Figure 4.33. This result is obtained for 20 rotation angles of the tilted loudspeaker array. Only a rotation around the vertical axis is used. The presented setup is applicable for a limited frequency range below $1\,\text{kHz}$ and for SH orders up to approx. $n = 5$.

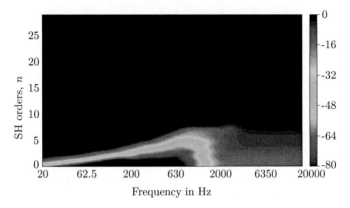

Figure 4.33.: Accumulated synthesis error in dB for SH orders over frequency for the dodecahedron system (Red indicating large errors, blue low errors) according to (J. KLEIN, 2012).

To increase the frequency limit more rotations to orientations different from the ones obtained by rotation around the vertical axis are required. The resolution of the free-field measurement was sufficiently high and much higher than $n = 5$ and hence not a limiting factor. Furthermore, the same membranes are used for all 12 channels. This leads to limited energy in specific SH orders and is addressed in the ongoing work of KLEIN (C. I. KLEIN, 2011; J. KLEIN et al., 2013).

Figure 4.34.: Cuboid target loudspeaker (left) and dodecahedron loudspeaker array (right) at the same source position in a small lecture room with omni-directional microphones and dummy head. All other loudspeakers shown were not used during the measurement.

4.5.3. Measurement in a Medium-sized Room

Figure 4.34 shows the measurement setup with the dodecahedron loudspeaker array and the rectangular cuboid target loudspeaker in the small lecture room at the ITA. The loudspeaker array and the target loudspeaker were both positioned on a turntable and measurements have been conducted every 27.7°.

The time variances during the measurement session are captured by the correlation coefficient of the measured impulse responses for different frequency bands referenced to an impulse response at around 30 minutes in Figure 4.35. In the beginning of the measurement session the room shows a settling time of approx. 5 to 10 minutes after the last person has left the room and closed the door. This time depends on the size of the room and on the maximum deviation in the correlation coefficient allowed. The effect of opening and closing the door to the room between two measurements in the range of $15 - 20$ min. and $50 - 55$ min. can be observed. Entering the room of a single person causes a drop of the correlation to approx. 96 % at 57 min. Typically, the temperature inside the room changes over time if the room is not air-conditioned. The influence of changes in temperature on the FRF are analyzed exemplarily in the appendix A.4 using again the analytic room model.

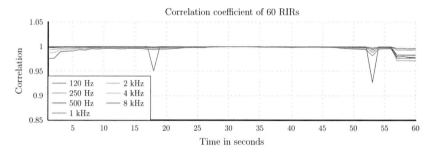

Figure 4.35.: Correlation coefficient of measured impulse responses to monitor time variances in a seminar room referenced to an impulse response in the center of the time scale for different frequency bands.

The impulse response and the frequency response measured with the target loudspeaker and the obtained by synthesis is presented exemplarily for an arbitrarily chosen position in Figure 4.36. All data available from the loudspeaker array measurements have been used for the synthesis. The order of magnitude in both time and frequency domain match and the fine structure shows also a good agreement.

The limitation of the method for the particular loudspeaker array and the orientations used during the measurement is depicted in Figure 4.37. In order to measure the similarity between the impulse responses the cross-correlation coefficient is calculated in third-octave bands. Different sets of input data for the synthesis are used characterized by the number of different orientation angles used. As can be seen, with the given setup the correlation coefficient is close to its theoretical maximum of 1 for frequencies in the range of 150 Hz up to 1 kHz. For lower frequencies the SNR decreases as the loudspeaker array as well as the target loudspeaker do not radiate sufficient sound power. Hence, the correlation coefficient theoretically decreases to its minimum value of 0 as two noise signals are compared. For frequencies above 1 kHz the correlation coefficient decreases as the method is not capable of synthesizing the correct impulse response due to reasons regarding the maximum order of spherical harmonics. This might be explained by the fact, that the maximum order of spherical harmonics required to describe a radiation pattern with a given quality rises over frequency.

Furthermore, the number of single impulse response measurements with the loudspeaker array must be higher than the number of spherical harmonic coefficients for the particular order. In addition, the radiation patterns of the rotated and/or tilted loudspeaker must provide sufficient energy for all spherical harmonic coefficients up the maximum order that is excited by the loudspeaker array. Summarizing these points, the method requires more and more measurements with the rotated and tilted loudspeaker array for an increasing upper frequency limit and the use of such a simple loudspeaker array as the dodecahedron setup is no longer suitable. Details on how to design a suitable loudspeaker array for the method are published compactly in (POLLOW et al., 2012) and in (J. KLEIN, 2012) in detail.

4.6. Application III—Directivity and Room Acoustic Parameters

The parametric model is used exemplarily for a virtual rotation of the cuboid target loudspeaker[16]. Findings of the influence of the radiation pattern of the sound source on room acoustic parameters and hence also a rotation of the sound source in a similar manner have been presented, e.g., in (MARTIN et al., 2007). The EDT and C_{80} have been chosen to prove the applicability of the proposed method for room acoustic uncertainty analysis as the radiation pattern can be

---

[16]Parts of the following results have been published in (DIETRICH et al., 2011; KUNKEMÖLLER et al., 2011)

clearly observed in these parameters by a simple step-wise rotation of the source. The results obtained by a real rotation of the target loudspeaker and a virtual rotation in the synthesis are compared in Figure 4.38. As can be seen, both parameters reflect the strong radiation pattern of the cuboid loudspeaker over the rotation of the source. The parameters are calculated in octave bands including noise subtraction in the evaluation. The synthesis agrees well with the original measurement in both shape over rotation angle and absolute value up to 1 kHz. However, small deviations can be observed that are assumed to be caused by a finite signal-to-noise-ratio and the fact that the correlation between measurement and synthesis is not perfect as shown before. However, the parameters also agree well for the frequency bands above 1 kHz although the correlation is lower for these frequencies. This can be explained by an averaging effect caused by the room acoustic parameter calculation. Nevertheless, the deviation of the room acoustic parameters of measurement and synthesis is expected to increase with decreasing correlation of the impulse responses. For the 2 kHz octave band the correlation coefficient indicates that simulation and real measurement do not agree. The room acoustic parameters in this frequency band also shows this mismatch.

This example shows that real-life measurement data with a specific measurement procedure can be used to simulate the uncertainty caused by the loudspeaker directivity. It has to be mentioned, that due to practical limitations mostly turntables are used to rotate the loudspeaker around its vertical axis. In order to investigate the uncertainty more thoroughly the loudspeaker should be moved towards arbitrary orientations. This could be done virtually with the presented method. However, as long as the measurement loudspeaker and the procedure are not capable of measuring up to a frequency band of 8 kHz the method is not able to predict the uncertainty for the entire frequency range specified in the standard. Research in this direction can be found in (J. KLEIN et al., 2013).

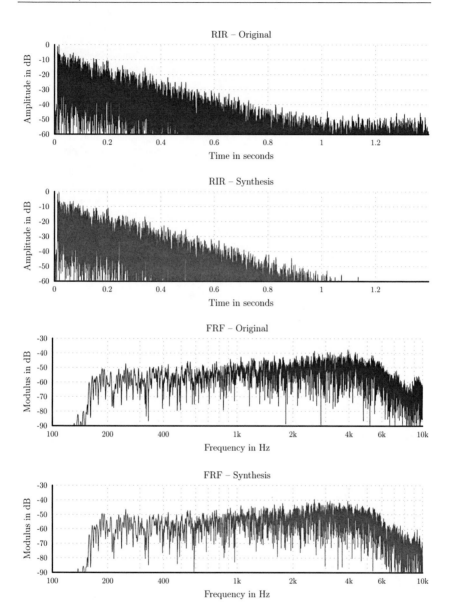

Figure 4.36.: Original (first row) RIR obtained with cuboid target loudspeaker and synthesis (second row) with the presented method using 240 individually measured RIRs and corresponding frequency response in the last two rows. A low-pass filter at 2 kHz has been applied for time plots.

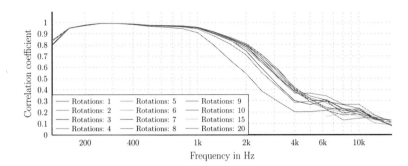

Figure 4.37.: Similarity between measured and synthesized room impulse response in terms of correlation coefficient over frequency for different sets of orientation angles used for the 12 channel loudspeaker array.

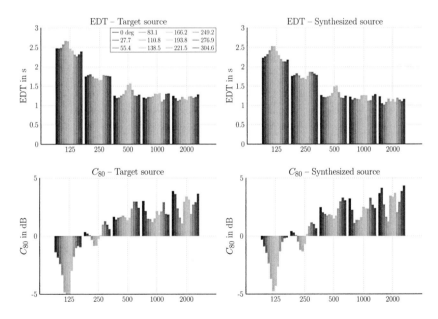

Figure 4.38.: Room acoustic parameters EDT and C_{80} for the measured and synthesized impulse responses over rotation angle of the cuboid target loudspeaker.

4.7. Summary and Scientific Contribution

Parametric analytical models for different applications in airborne sound transmission were presented. The influence of the precision in the positioning of microphones was studied as well as the influence of the directivity of a loudspeaker and its orientation.

In the first application example from the field of in-situ sound barrier measurements an array of microphones is used to capture an angle-averaged reflection index. An analytical model was developed that describes the sound propagation for a directive sound source to the microphones and additionally the reflection at the barrier. The loudspeaker's directivity is modeled by a cap model using spherical harmonics. The uncertainty of the position of the array microphones and the uncertainty in the orientation of the directive loudspeaker has been estimated based on empiric data from laboratory and round robin measurements. These values were used as uncertain input quantities for the parametric model along with Monte Carlo simulations. The simulation results describe the uncertainty of this reflection index for different accuracy in positioning of the microphones and the loudspeaker separately. The influence of the loudspeaker directivity on the uncertainty was found at high frequencies above approx. 2 kHz for the specific size of the chosen loudspeaker. Furthermore, the results show an increased uncertainty for the low frequency bands below approx. 200 Hz also caused by the directivity at high frequencies in conjunction with the complex evaluation procedure for the reflection index. Especially low reflection factors are associated with a higher measurement uncertainty.

For the second application example from the field of room acoustics, the link between an analytic model for rectangular rooms and measurement results has been shown using a rational fit algorithm. The agreement of this fit is promising for the low frequency range up to a frequency of approx. $400 - 500$ Hz in large rooms due to a limit caused by the increase of the modal density. The analytic model was further used to investigate deviations in the modal coefficients with increasing deviations in the position of the receiver. These effects in the modal coefficients were shown to be related to the wavelength and the modal density. Uncertainties in the room acoustic parameters caused by uncertain positions were analyzed and found to be in good agreement with the measurements. For distances of approx. $^1/_2$ of a wavelength the uncertainty in the modal coefficients as well as the room acoustic parameters have been found to reach a constant value.

The enhancement of the analytic model towards arbitrary source and receiver characteristics was used to investigate the influence of the rotation of dipoles and quadrupoles on the modal coefficients. The derivation of room acoustic parameters yielded a relationship between the uncertainty of the orientation of these sources with basic radiation patterns and the uncertainty in the room acoustic parameters. The suitable frequency range for this modal approach is limited towards higher frequencies as the modal density rises quadratically and the calculation time increases accordingly.

The third example involves a measurement method developed for room impulse responses with arbitrary radiation patterns. The neccessary measurement setup and especially the loudspeaker array used limit the maximum frequency to approx. 1 kHz. However, the derivation of room acoustic parameters obtained from a target source with distinct radiation pattern and a synthesis by the method showed very good agreement up to this frequency. This method is expected to become a powerful tool for uncertainty analysis once a suitable loudspeaker array is manufactured and hence capable of yielding results up to at least $4 - 5$ kHz. These data can then be used to analyze the effects and deviation caused by different commercial loudspeakers for room acoustic measurements but with the advantage of using the same room characteristics and the exact same position in a virtual measurement. However, the combination of the analytic room model including radiation patterns with this measurement method in combination with realistic radiation patterns of loudspeakers is expected to provide more insight into the applicability of the measurement method and finally also on the measurement uncertainty in room acoustic parameters.

The examples investigated capture a wide range and they are assumed to be transferable to different measurements of airborne transfer paths.

5

Uncertainties in Structure-borne Transfer Paths

This chapter deals with the principal concept of characterizing structure-borne sound sources and their interaction with the connected structure. A matrix notation based on the general case is deduced for three Transfer Path Analysis (TPA) methods and enhanced by further consideration of vibration isolators. Based on this notation the relationship between the TPA methods is investigated and the principal differences are discussed. The uncertainty in the mobility caused by inaccuracies in positioning of the sensors is studied by means of Monte Carlo simulations using an analytic model for rectangular plates as a first application example. Moreover, this model and the general notation is used in a second example investigating the deviations caused by applying typical simplifications to the measurement setup.

5.1. Remark on Structure-borne Uncertainty Analysis

"The estimation of structure-borne sound power is generally thought to be a 'delicate' problem, meaning that the output quantity (transmitted power), may vary significantly due to apparently small changes in the source or receiver structures or the details of the connections." (Evans, 2010)

The sources of uncertainty in structure-borne sound modeling and the estimated sound power have already been studied in detail by Evans and summarized in 8 classes:

1. Variability from one structure to another (inter-variability);

2. Variability of one structure with time (intra-variability);

3. Variability in the analysis or measurement process;

4. Uncertainty in geometrical and physical parameters;

5. Uncertainty due to assumptions and approximations in the modeling of a structure;

6. Uncertainty in measurement;

7. Uncertainty due to granularity in the input data of a structure-borne sound power estimate;

8. Uncertainty due to source characterization methods.

In the following, the focus is therefore set to the principal equations in a more theoretical perspective. The points 4 and 5 of this list are touched by the following investigation only.

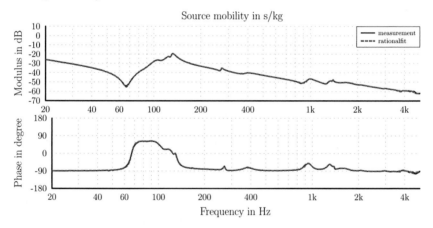

Figure 5.1.: Measured source mobility of an Visaton EX 60 S vibration exciter in normal direction and rational fit with 35 poles as an example to show the principal applicability of the modal modeling approach for structure-borne scenario.

In general, the modal approach as investigated for the airborne scenario is applicable for structure-borne systems as well (F. FAHY and GARDONIO, 2007). In order to provide a practical example, the driving point mobility in normal direction of an vibration exciter (Visaton EX 60 S) measured in free condition is depicted in Figure 5.1. As can be seen, the transfer function between force and velocity contains resonances and is further minimum-phase. Furthermore, this mobility is fitted by the approach introduced in Section 2.3.3 and already applied successfully to airborne transfer functions in Section 4.3.3. The number of resonances used for the rational fit algorithm is 35. The results is found in very good agreement with the measurement data for both magnitude and phase. Compared to the room transfer functions the modal density is much lower in this structure-borne example. Nevertheless, the applicability of the modal approach enables a similar modeling of the uncertainties with varying geometry, temperature or position and orientation as in the airborne scenario. However, the structure borne sound transmission strongly depends on the source-receiver interaction which was not considered in the airborne scenario. Furthermore, the comparable field quantity to the structure-borne force vector is the scalar sound pressure, leading to the consideration of more than one degree of freedom instead. Hence, these two differences and especially their influence on the uncertainty are investigated in the following.

5.2. Source, Receiver and their Interaction

For the airborne sound the coupling between the source and the transfer path or the medium can be reasonably neglected since the acoustic radiation impedance and the source impedance are commonly not in the same order of magnitude for, e.g., room acoustic application scenarios. When considering structure-borne sound propagation this coupling can only be neglected in some special cases in more detail. But most transfer path measurements practiced implicitly assume these cases. Furthermore, the fact that structure-borne sound sources in general have *6 degrees of freedom* as well as the transfer paths is neglected. Then only the translation component in normal direction to the structure is accounted. But also the other components can have a significant contribution as shown in (PETERSSON and GIBBS, 2000). The practitioner's goal is to find a suitable trade-off between granularity and accuracy (EVANS, 2010).

The ideal approach to a suitable modeling of a structure-borne scenario therefore accounts for a full description and measurement of the source. In a next step, by applying reasonable assumptions the model could be reduced by neglecting minor contributions. Furthermore, the impedances or mobilities describing the coupling situation of the source and receiver have to be measured. Commonly, this approach is not applicable in practice due to time and cost constraints. Furthermore, the contact is commonly also assumed to be an infinitesimal point contact. But there are various applications where a multi-point contact or a contact over an area is more realistic. This becomes evident for short wavelengths and hence high frequencies when the contact area cannot be considered as small compared to the wavelength. The interaction between the contact points and the reaction on the source can also be investigated in more detail. In addition, the transfer paths and impedances have to be considered for the 6 degrees of freedom separately. Only if the propagation is entirely described reasonable assumptions can be applied (PETERSSON and GIBBS, 2000). Point contacts are assumed in the following.

5.2.1. Impedance and Mobility

The relation between the force vector and the velocity vector in one contact point is described by the point mobility matrix $\mathbf{Y}$ as

$$\mathbf{v} = \mathbf{Y} \cdot \mathbf{F}. \tag{5.1}$$

The velocity is considered as a weighted answer to the causing force (F. FAHY and GARDONIO, 2007). The mobility can be considered as a frequency dependent transfer function between the force signal and the velocity signal and hence all mobilities are causal. The mobility component in the matrix connecting the force or moment with the velocity or angular velocity of the same degree of freedom (diagonal of the matrix) is called *driving-point mobility* and it is furthermore *minimum-phase* according to (F. FAHY and GARDONIO, 2007). For real and passive systems the phase of these components can be further narrowed to lie between $\pm 90°$ which is required to fulfill the conservation of energy. The remaining components are *transfer mobilities* and are in general not minimum-phase. Hence, the phase generally can exceed this range.

The impedance is defined as the inverse of the mobility. For one degree of freedom the point impedance is hence also minimum-phase (TOHYAMA and KOIKE, 1998) and hence also causal. It is important to mention that the impedance matrix can only be obtained by inverting the entire mobility matrix. The transfer impedances can be a-causal. Both matrices are quadratic and can be inverted in general despite the problems due to noise or problems due to the very high condition numbers.

In order to suppress influences by measurement noise the force and velocity impulse response obtained either by impact hammer or shaker measurements can be time windowed with the techniques described in Section 2.2. These impulse responses are causal, as they always follow the excitation signal. With the knowledge that mobilities are causal impulse responses between force and velocity they can be further time windowed after applying the spectral division as proposed in (DIETRICH and LIEVENS, 2009; DIETRICH, LIEVENS, and PAUL, 2009).

Symmetry of Mobility Matrices   The mobility matrix at one contact point can be considered to have a general structure

$$\mathbf{Y}_{\mathrm{el}} = \begin{pmatrix} \mathbf{Y}_{\mathrm{TT}} & \mathbf{Y}_{\mathrm{TR}} \\ \mathbf{Y}_{\mathrm{RT}} & \mathbf{Y}_{\mathrm{RR}} \end{pmatrix} \tag{5.2}$$

where the subscript T denotes the translational and R denotes the rotational components. The entries on the diagonal of this block matrix consider the inter-point mobilities of each degree of freedom and the interaction between all rotational components and the interaction among all rotational components.

For a single structure-borne sound source i with N contact points the entire mobility matrix reads as

$$\mathbf{Y}_{\mathrm{s},i} = \begin{pmatrix} \mathbf{Y}_{\mathrm{el},11} & \cdots & \mathbf{Y}_{\mathrm{el},1N} \\ \vdots & & \vdots \\ \mathbf{Y}_{\mathrm{el},N1} & \cdots & \mathbf{Y}_{\mathrm{el},NN} \end{pmatrix} \tag{5.3}$$

and finally for L independent sound sources as

$$\mathbf{Y}_{\mathrm{s}} = \mathrm{diag}(\mathbf{Y}_{\mathrm{el},1} \ldots \mathbf{Y}_{\mathrm{el},L}). \tag{5.4}$$

The independent sound sources are assumed to have no connection among each if they are not connected to the structure. Hence, the coupling between these sources vanishes. On the other hand, the mobility matrix of the structure $\mathbf{Y}_{\mathrm{r}}$ is not such a sparse matrix and hence the mobility matrix of the coupled system is also not sparse.

The mobility matrix on one contact point i is furthermore symmetric according to the Raleigh reciprocity theorem (F. J. FAHY, 1995)

$$\mathbf{Y}_{\mathrm{el},ii} = \mathbf{Y}_{\mathrm{el},ii}^{T}. \tag{5.5}$$

The same holds also for the cross-coupling sub-matrices

$$\mathbf{Y}_{\mathrm{el},ij} = \mathbf{Y}_{\mathrm{el},ij}^{T}. \tag{5.6}$$

Furthermore, the cross-coupling sub-matrices have a symmetry among each others

$$\mathbf{Y}_{\mathrm{el},ij} = \mathbf{Y}_{\mathrm{el},ji}^{T}. \tag{5.7}$$

In summary, the mobility matrix is symmetric and has additional internal symmetry. This symmetry can be utilized to cross-check measurements if all components are measured or it can be used to reduce the number of measurements.

5.2.2. Characterization of Structure-borne Sound Sources

Theoretically, a structure-borne sound source can be entirely characterized by either its blocked force and its free velocity and its mobility matrix including all connection points if it can be reasonably considered as an Linear Time-Invariant (LTI) system. Depending on the type of source and further constraints for possible mountings of the source the measurement of either free velocity or blocked force is preferable. The measurement of the blocked force is conducted by mounting the source to a structure with zero mobility. In practice, a very low mobility in relation to the source mobility is chosen instead.

Many publications consider a general source characterization (MONDOT and A. T. MOORHOUSE, 2000; MONDOT and PETERSSON, 1987; A. T. MOORHOUSE, 2001; A. MOORHOUSE, 2007; A. T. MOORHOUSE and GIBBS, 1995). The approach of directly applying fully determined impedance or admittance matrices for all degrees of freedom in a multi-point contact situation leads to numerical and algorithmic problems[1]. Therefore PETERSSON UND GIBBS (PETERSSON and GIBBS, 2000) try to avoid this be introducing an effective mobility Y,

$$Y_{ii}^{nn\Sigma} = Y_{ii}^{nn} + \sum_{j=1,j\neq i}^{6} Y_{ii}^{nn} \frac{F_j^k}{F_i^n} + \sum_{k=1,k\neq n}^{N} Y_{ii}^{nk} \frac{F_i^k}{F_i^n} + \sum_{k=1,k\neq n}^{N} \sum_{j=1,j\neq i}^{6} Y_{ij}^{nk} \frac{F_j^k}{F_i^n}. \tag{5.8}$$

This approach combines the amplitude of the force of each degree of freedom to obtain a weighted mean of the mobility in one contact summarizing all degrees of freedom. The first term expresses the point mobility, the second term the coupling of various components (degrees of freedom) in one contact point (intra-point cross-coupling) and the last term the coupling of the point among each others (inter-point cross-coupling). FAHY UND GIBBS emphasize that only one fourth is

---

[1] Full matrices are used in the following and the condition number is considered as well. For the theoretic cases studied in these thesis these problems are negligible but become problematic in real-life measurements especially with finite Signal-to-Noise Ratio (SNR).

related to the translation components only and the remaining terms are related to the rotation components as well which is commonly underestimated in practice[2].

The blocked force $\mathbf{F}_b$ can be transferred to the free velocity $\mathbf{v}_f$ of the source as

$$\mathbf{v}_f = \mathbf{Y}_s\mathbf{F}_b. \tag{5.9}$$

5.2.3. Modeling the Coupling of Source and Receiver

The coupling function is dependent on the source and the receiver mobility and can be explained using circuit diagrams as commonly used in electrical engineering as, e.g., depicted in Figure 5.2. The velocity and the force at the interface points to the structure linearly scale with the source signal but also scale with a combination of the mobilities.

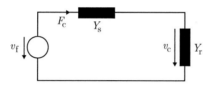

Figure 5.2.: Equivalent circuit diagram with source and receiver mobility. The source signal is represented by the free velocity. The voltage corresponds to the velocity and the current corresponds to the force.

The mobility matrix of the coupled matrix $\mathbf{Y}_c$ is deduced and the analogy that substitutes voltage by velocity is used. The free velocity is connected to the force in the coupled condition $\mathbf{F}_c$ according to

$$\mathbf{F}_c = \left(\mathbf{Y}_s + \mathbf{Y}_r\right)^{-1} \cdot \mathbf{v}_f = \left(\mathbf{Y}_s + \mathbf{Y}_r\right)^{-1} \cdot \mathbf{Y}_s \cdot \mathbf{F}_b. \tag{5.10}$$

The velocity in the coupled condition $\mathbf{v}_c$ can hence be derived as

$$\mathbf{v}_c = \mathbf{Y}_r \cdot \mathbf{F}_c = \mathbf{Y}_r \cdot \left(\mathbf{Y}_s + \mathbf{Y}_r\right)^{-1} \cdot \mathbf{Y}_s \cdot \mathbf{F}_b. \tag{5.11}$$

By comparison with Eq. 5.1 the mobility of the coupled system follows as

$$\mathbf{Y}_c = \mathbf{Y}_r \cdot \left(\mathbf{Y}_s + \mathbf{Y}_r\right)^{-1}\mathbf{Y}_s. \tag{5.12}$$

---

[2]It has to be mentioned, that a simple comparison of the magnitude of the force spectrum for all degrees of freedom is not sufficient to decide whether this path is important to consider. At this point the combination of the source signal and the transfer path including the coupling function yielding the path contribution has to be considered.

From linear algebra the following formulations are equivalent

$$\mathbf{Y}_c = (\mathbf{Z}_r + \mathbf{Z}_s)^{-1} = (\mathbf{Y}_r^{-1} + \mathbf{Y}_s^{-1})^{-1} = (\mathbf{Y}_s^{-1} + \mathbf{Y}_r^{-1})^{-1}. \qquad (5.13)$$

Due to reasons of symmetry the matrices $\mathbf{Y}_s$ and $\mathbf{Y}_r$ can also be exchanged, which can also be explained by using the associative law and Eq. 5.13. It has been shown that formulations using only a single inversion can be advantageous (HOELLER, 2010; LIEVENS, 2010).

The measurement setup to measure the blocked force uses a structure with $\mathbf{Y}_s$ tending to zero rigidly coupled to the source and the forces at the contacts points are measured. This setup can be explained by inserting $\mathbf{Y}_s = 0$ in Eq. 5.10 and it follows $\mathbf{F}_c = \mathbf{F}_b$ for this particular case.

5.3. TPA Methods in Matrix Notation

TPA methods can be divided into *load-response* and *response-response* methods (P. A. GAJDATSY, 2011). The first uses the measured force as the cause of the measured velocity. The latter uses velocity measurements only[3]. In general, the methods can be also classified due to the condition the components are measured in as:

- Coupled—Sources connected to the structure (simple measurements, mostly less flexibility regarding the use of the results);

- Decoupled—Physical separation of sources and receiver (extensive procedure and extensive measurements, more flexibility).

By consideration of the measurement methods in terms of the matrix equations arbitrary numbers of source, degrees of freedom and contact points can be investigated. The equations of three different typical TPA methods are deduced and the relation of the results obtained by these methods is presented. The effect of cross-coupling on the results and their difference can be observed by this approach[4].

---

[3]The acceleration is measured in practice instead. However, all formulations in the following use the velocity.

[4]Parts of the following have already been submitted in (LIEVENS et al., (2013)) and are therefore presented in (LIEVENS, 2013).

The introduction of such a matrix notation is not entirely new, but a unified notation is used in this thesis to connect the different methods. E.g., MOOR-HOUSE also uses a matrix notation but he does not consider the TPA methods. The detailed overview of TPA methods is presented, e.g., in (LMS, 1995; LMS, 2007; VAN DER AUWERAER et al., 2007) but without a matrix notation and hence no direct comparison of the results from a theoretical point of view.

5.3.1. Load-Response Methods

Two different load-response methods are investigated in the following. Firstly, a frequently used method that involves measurements of the decoupled system. It is therefore denoted as *classic TPA* in this thesis. Secondly, a measurement method of the coupled system.

Classic TPA

The classic TPA uses measurements of the decoupled sources and transfer path measurements of the structure with all sources being removed. On the one side, the source is characterized as explained above. On the other side, the structure is measured by introducing a force at each connection point and measuring the responding velocities of the receiver and the sound pressure. The velocities are normalized to the measured force and yield the mobility. The sound pressure is also normalized to the force and yields the transfer path $\mathbf{H_r}$. A model of the source using internal sources that itself involve transfer paths to the connection points and the coupling of the source to the receiver is illustrated schematically in Figure 5.3.

Hence, the required measurements are in summary:

- Blocked force $\mathbf{F_b}$ (active, running source);

- Mobility $\mathbf{Y_s}$ (measurement of F and v, ext. excitation) at the idle source;

- Mobility $\mathbf{Y_r}$ and transfer path $\mathbf{H_r}$ (measurement of F, v, p, ext. excitation).

The sound pressure at the receiver position can then be synthesized by using the transfer path obtained in decoupled condition $\mathbf{H_r}$ and the force in coupled condition $\mathbf{F_c}$ as

$$p = \mathbf{H_r} \cdot \mathbf{F_c}. \qquad (5.14)$$

where $\mathbf{F_c}$ is usually obtained by using Eq. 5.10 yielding

$$p = \mathbf{H_r}(\mathbf{Y_s} + \mathbf{Y_r})^{-1}\mathbf{Y_s}\mathbf{F_b} = \mathbf{H_r}\mathbf{CF_b} \qquad (5.15)$$

where $\mathbf{C}$ describes the coupling. This forward synthesis from the source to the receiving points is the Transfer Path Synthesis (TPS). This dimensionless matrix $\mathbf{C}$ describes the coupling between the source and the receiver, the coupling between the degrees of freedom and also the coupling between the interfaces. The errors introduced by applying simplifications or assumption and hence neglecting components of this matrix or setting them to unity

Furthermore, the velocities at these contact points can be obtained as

$$\mathbf{v_c} = \mathbf{Y_r}(\mathbf{Y_s} + \mathbf{Y_r})^{-1}\mathbf{Y_s}\mathbf{F_b} = \mathbf{Y_c}\mathbf{F_b} = \mathbf{Y_r}\mathbf{F_c}. \qquad (5.16)$$

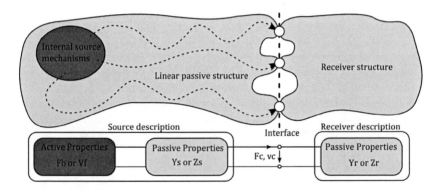

Figure 5.3.: Model of a structure-borne sound source with internal source components and connection to the receiver structure and block diagram after (HOELLER, 2010).

Coupled System

The measurement of transfer paths can also be realized using the coupled system. This does not involve the troublesome decoupling of the sources, but—as a drawback—does also not deliver information on the source signals. The transfer paths could, e.g., be measured by using an impact hammer at approximately the connection points. This force introduced externally is equivalent to the blocked force described above. Additionally, mounting of force sensors between the source and the receiver yields the same transfer paths but has the further advantage of measuring the in-situ force signals $\mathbf{F}_c$. The relation of the transfer path obtained by this method with the transfer path obtained by the classical TPA is of interest.

In the same manner as in Eq. 5.14 the relation between the introduced force and the sound pressure can be written as

$$p = \mathbf{H}_c \cdot \mathbf{F}_{ext} = \mathbf{H}_c \cdot \mathbf{F}_b \tag{5.17}$$

and by using Eq. 5.14, Eq. 5.15 and Eq. 5.17

$$p = \underbrace{\mathbf{H}_r(\mathbf{Y}_s + \mathbf{Y}_r)^{-1}\mathbf{Y}_s}_{\mathbf{H}_c} \mathbf{F}_b = \mathbf{H}_r \underbrace{(\mathbf{Y}_s + \mathbf{Y}_r)^{-1}\mathbf{Y}_s \mathbf{F}_b}_{\mathbf{F}_c} = \mathbf{H}_r \mathbf{F}_c = \mathbf{H}_c \mathbf{F}_b \tag{5.18}$$

with

$$\mathbf{H}_c = \mathbf{H}_r(\mathbf{Y}_s + \mathbf{Y}_r)^{-1}\mathbf{Y}_s. \tag{5.19}$$

$\mathbf{H}_c$ is the transfer path of the coupled system as also used in (ELLIOTT and A. MOORHOUSE, 2010) and (LIEVENS, 2010) measured with external excitation at the coupling points. This introduction of an external force is equivalent to a structure-borne sound source with this same internal force in terms of a blocked forced description.

It is important to mention that the transfer paths obtained from a measurement of a decoupled system and a coupled system are not identical according to Eq. 5.19. If the source mobility can be considered much greater than the receiver mobility, $\mathbf{H}_c$ approximates $\mathbf{H}_r$. Additionally, this constraint has to be fulfilled for all corresponding entries in both matrices. By taking $\mathbf{H}_r$ as the reference high uncertainties are to be expected when the assumption is not valid and hence the mobilities would be in the same order of magnitude. In case the source mobility is much lower than the receiver mobility the approximation is not applicable. Further measurements of $\mathbf{Y}_s$ and $\mathbf{Y}_s$ are required instead increasing again the measurement time and effort if they have not been measured before.

5.3.2. Response-Response Methods

Only one response-response method is chosen exemplarily. The Operational Transfer Path Analysis (OTPA) involves only velocity-based measurements (response) of the coupled system in running condition. It therefore eliminates the time consuming decoupling process and does not require force measurements (loads). The method was initially claimed to deliver quick and reliable results to find dominant and hence problematic transfer paths (NOUMURA and YOSHIDA, 2006a; NOUMURA and YOSHIDA, 2006b). It involves the *MIMO technique* (DE KLERK and OSSIPOV, 2010).

It is of interest in the scope of this thesis to investigate in how far the obtained transfer paths correspond to the results from the classical TPA. From a practitioner's point of view, it might be also interesting to investigate in how far the path contributions obtained by this method can be used as a measure for dominant paths. This question as well as the aspects of the Multiple Input Multiple Output (MIMO) technique and further negligence of transfer paths, correlation of internal source signals is addressed in detail in (P. GAJDATSY, 2008a; P. GAJDATSY, 2008b; P. A. GAJDATSY, 2011; P. GAJDATSY et al., 2011; P. GAJDATSY et al., 2008; P. GAJDATSY et al., 2011). Motivated by the practical problems—mainly due to condition numbers, noise and unintended negligence of paths—observed the *OPAX* method was developed (JANSSENS et al., 2011). This method involves additional measurements in coupled condition and an approach similar to the fitting described before to overcome these problems.

It is not important for the following equations how the transfer paths are obtained in terms of signal-processing methods, e.g., by MIMO techniques or if correlation of the source signals was present. It is assumed that the transfer path as described by the method is measured ideally without further artifacts.

The transfer path obtained by this method is denoted as $\mathbf{H}_a$. This transfer path captures a relation between the velocity at the interfaces—originally the method uses the acceleration a instead—and the sound pressure as

$$p = \mathbf{H}_a \mathbf{v}_\mathrm{c}. \tag{5.20}$$

According to Eq. 5.16 to relation to classical TPA using $\mathbf{F}_\mathrm{c}$ can be found

$$p = \mathbf{H}_a \mathbf{Y}_\mathrm{r} \mathbf{F}_\mathrm{c} \tag{5.21}$$

and comparison with Eq. 5.14 finally yields

$$\mathbf{H_r} = \mathbf{H}_a \mathbf{Y_r}. \tag{5.22}$$

The obtained transfer path is not identical to the result obtained by the classical TPA. Especially the physical units in $\mathbf{H}_a$ differ from $\mathbf{H_r}$. But in case $\mathbf{Y_r}$ can be reasonably assumed to be a diagonal matrix, the transfer paths for the different contact points can be compared by further claiming that the entries of $\mathbf{Y_r}$ to be in the same order of magnitude. This assumption is only based on $\mathbf{Y_r}$ and not on both mobilities of the source and the receiver.

In order to overcome the problems of the MIMO technique a combination with coupled TPA measurements can be used. The transfer paths in coupled condition $\mathbf{H_c}$ are not subject to these problems and are in known relation to $\mathbf{H_r}$. Right multiplication of the inverse matrices on the right hand side in Eq. 5.19 yield

$$\mathbf{H_c}\mathbf{Y_s}^{-1}(\mathbf{Y_r} + \mathbf{Y_s}) = \mathbf{H_r}. \tag{5.23}$$

Inserting into Eq. 5.22 and right multiplication by the inverse of $\mathbf{Y_r}$ yields

$$\mathbf{H_c} = \mathbf{H}_a \mathbf{Y_r}(\mathbf{Y_r} + \mathbf{Y_s})^{-1}\mathbf{Y_s} = \mathbf{H}_a\mathbf{Y_c} \tag{5.24}$$

and finally the relation to the transfer path obtained by the coupled TPA can be written as

$$\mathbf{H}_a = \mathbf{H_c}\mathbf{Y_c}^{-1}. \tag{5.25}$$

Only the measurement of the coupled mobility and hence no decoupling of the sources is required.

5.3.3. Relation of the Methods and Path Contributions

The relation of the methods as deduced above are in summary

$$
\begin{aligned}
p_{\text{classic TPA}} &= \mathbf{H_r} & \cdot \mathbf{F_c} = \mathbf{H_r} & \cdot \mathbf{F_c} \\
p_{\text{coupled TPA}} &= \mathbf{H_r}(\mathbf{Y_s} + \mathbf{Y_r})^{-1}\mathbf{Y_s}\cdot \mathbf{F_b} = \mathbf{H_c}\mathbf{Y_s}^{-1}(\mathbf{Y_r} + \mathbf{Y_s}) \cdot \mathbf{F_c} \\
p_{\text{OTPA}} &= \mathbf{H_r}\mathbf{Y_r}^{-1} & \cdot \mathbf{v_c} = \mathbf{H}_a\mathbf{Y_r} & \cdot \mathbf{F_c}.
\end{aligned} \tag{5.26}
$$

For the path contributions the vector to the right hand-side in the center column describing the source signal is substituted by a diagonal matrix, containing the values of the vector of the same rows. Ideally, all sound pressures on the left hand-side are identical if no simplifications are applied and the measurements results are assumed to be ideal with uncertainty but the path contributions are not identical. Figure 5.4 illustrates a typical plot of the path contribution in a TPA. The contributions are shown separately for the contact interfaces considered for the OTPA and the classical TPA method summarized in frequency bands. The results are exemplary calculated by using an implemented simulator as described in the appendix Section A.7. The OTPA result was calculated based on simulations of $\mathbf{H_r}$, $\mathbf{Y_r}$, $\mathbf{Y_s}$ and a given force vector. The path contribution for the given example are generally similar, but specific deviations can be observed especially for the 155 Hz and 200 Hz frequency band. The classical TPA method indicates that all 9 paths contribute almost equally, but the OTPA indicates the paths 2, 5 and 9 to be the problematic paths. The sum of the path contributions for each of the methods are identical, only the weighting of the importance of the paths is different.

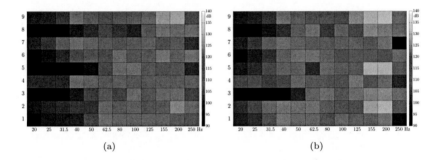

(a) (b)

Figure 5.4.: Path contribution (color indicating level in dB) for 9 different transfer paths obtained by the classical TPA method (left) and theoretical results obtained by the OTPA method (right).

5.4. Characterization of Vibration Isolators

Vibration isolators are used to attenuate the structure-borne sound propagation from the source to the receiving structure. Fundamentals in modeling of such isolators is presented in the following assuming linearity. This formulation is then adapted to the matrix notation compatible to the matrix formulation deduced for the TPA methods.

5.4.1. Two-Port Theory and Modeling

When it comes to the description of transfer elements, two-ports are advantageous and commonly applied in electrical engineering (FELDTKELLER, 1944). In general, a two-port can, e.g., be fully characterized by its impedance matrix or by its transfer matrix. Therefore, it contains all information on the input, output and transfer impedances from input to output of the system and vice versa. The use of two-ports for vibration problems has been proposed by (MOLLOY, 1957) by applying the methods from electrical engineering. For applications in automotive industry this two-port concept has also been integrated for the binaural transfer path analysis (SOTTEK, 2006). It is of further interest how vibration isolators could be introduced in terms of the matrix notation. The formulation in the following as also found in the literature is valid for the one-dimensional case.

Vibrational isolators are assumed to be linear in the following which is valid for small amplitudes and if the pre-load does not chance significantly. These limitations and the problems occurring for a broad frequency range have been investigated in (VERMEULEN, LEMMEN, and VERHEIJ, 2001).

The standard *ISO 10846* (ISO 10846, 2005) describes the measurement of vibration isolators. It mainly uses an indirect measurement method. DICKENS further investigates the modeling, measurement and the prediction of sound propagation through these isolators (DICKENS, 1998; DICKENS, 2002; DICKENS and NORWOOD, 2001). He mainly proposes a direct measurement method and investigates the limitations of this measurement approach. The measurement setup compromises the accuracy as flanking transmission in the setup increases over frequency and the influence on the measurement result increases accordingly.

The theoretical description of the sound propagation via a single vibration isolator by means of two-ports uses a 2×2 matrix notation for the one-dimensional case. The notation already has similarities with the TPA notation but the TPA notation uses only one element for a one-dimensional transmission instead of four. The notation in the following links the force F_1 and the velocity v_1 at the input of the isolator with the force F_2 and velocity v_2 at its output.

Mobility or Impedance Matrix

The impedance matrix is often used to describe isolators as meaningful quantities, e.g., the input and the output impedance are the entries on the diagonal of the matrix. Remaining elements are the transfer impedances between input and output:

$$\begin{pmatrix} F_1 \\ F_2 \end{pmatrix} = \begin{pmatrix} Z_{11} & Z_{12} \\ Z_{21} & Z_{22} \end{pmatrix} \begin{pmatrix} v_1 \\ v_2 \end{pmatrix} = \mathbf{Z} \begin{pmatrix} v_1 \\ v_2 \end{pmatrix}. \tag{5.27}$$

The determination of the elements can be separated as

$$Z_{11} = \frac{F_1}{v_1}\bigg|_{F_2=0} \qquad Z_{12} = \frac{F_1}{v_2}\bigg|_{v_1=0} \tag{5.28}$$

and

$$Z_{21} = \frac{F_2}{v_1}\bigg|_{v_2=0} \qquad Z_{22} = \frac{F_2}{v_2}\bigg|_{F_1=0} \tag{5.29}$$

where the isolator has to be connected to two different termination impedances—ideally free $(F = 0)$ and blocked $(v = 0)$—at its input and its output.

The input and output impedance of an isolator connected to the source on the one hand and to the receiver on the other hand is calculated as

$$Z_{\text{in}} = Z_{11} - \frac{Z_{12}Z_{21}}{Z_{22} + Z_{\text{r}}}$$
$$Z_{\text{out}} = Z_{22} - \frac{Z_{12}Z_{21}}{Z_{11} + Z_{\text{s}}} \tag{5.30}$$

with the source impedance Z_{s} and the receiver impedance Z_{r} (DICKENS, 1998). These values represent the impedance of the structure seen by the source through the isolator and vice versa. The mobility or admittance matrix is defined analog to the impedance matrix. It has to be pointed out, that the entries of the matrix can become infinite for theoretical cases. Hence, the contrary formulation, e.g. impedance instead of mobility, might become beneficial as the infinite values become zeros instead. These formulations are not directly applicable for multi-ports.

Transmission Matrix

The transmission matrix $\mathbf{T}$ (or $ABCD$ matrix) is mainly applied in the field of physics and especially optics. This can be explained as a series connection of transfer elements can be calculated by matrix multiplication of the transmission matrices. This is not directly possible with the impedance matrix. The relation between the input and output reads as

$$\begin{pmatrix} F_1 \\ v_1 \end{pmatrix} = \begin{pmatrix} A & B \\ C & D \end{pmatrix} \begin{pmatrix} F_2 \\ v_2 \end{pmatrix} = \mathbf{T} \cdot \begin{pmatrix} F_2 \\ v_2 \end{pmatrix} \tag{5.31}$$

where the elements of the matrix are determined with the following constraints

$$\begin{aligned} A &= \left.\frac{F_1}{F_2}\right|_{v_2=0} & B &= \left.\frac{F_1}{v_2}\right|_{F_2=0} \\ C &= \left.\frac{v_1}{F_2}\right|_{v_2=0} & D &= \left.\frac{v_1}{v_2}\right|_{F_2=0} \end{aligned} \tag{5.32}$$

also measured with free and blocked condition.

Both transmission matrix and impedance or mobility matrix representation can be transformed into each other as, e.g., described in (HYNNÄ, 2002).

Typical Simplifications for Measurements

Based on the geometrical symmetry of the isolator simplifications to the general case involving 6 degrees of freedom can be applied. In case of rotational symmetry of the isolation elements simplifications can be applied as some elements in the matrix can be assumed to vanish. Furthermore, elements are redundant and hence measurements can be reasonably simplified. If the isolators are also symmetric at the input and output further simplifications become possible. By assuming no intra-point cross-coupling, again only the elements on the diagonal are important. This is summarized in *ISO 10846*. According to the standard only two or three elements of the matrix are sufficient in many cases to describe the isolator.

As also mentioned for the transfer paths before, the dominance of one degree of freedom is not only depending on the magnitude of the elements in this matrix but also on the magnitude of the driving signals. Hence, these simplifications can only be applied reasonably for a specific scenario and require detailed experience.

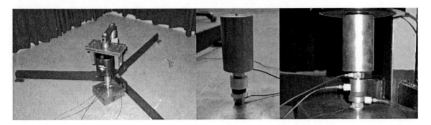

Figure 5.5.: Exemplary measurement setup for the characterization of small vibration isolators in normal direction with a pre-load mass and a shaker as used in refrigerators to decouple the compressor from the housing.

Isolators with symmetry regarding the input and output can be described by only two parameters for each frequency bin. One single measurement in blocked condition is sufficient. By using the *Raleigh reciprocity theorem* the remaining parameters can be obtained. All passive isolators are subject to the reciprocity theorem and it holds $Z_{12} = Z_{21}$ Furthermore, the input and output impedance for symmetric isolators are equal $Z_{11} = Z_{22}$. The symmetry of the transmission matrix follows accordingly. A measurement of a symmetric vibration isolator in normal direction as, e.g., applied to decouple a compressor from the refrigerator is depicted in Figure 5.5. To capture the influence of the weight of the compressor on the soft isolator a pre-load is applied using an appropriate mass driven by the shaker is used (DIETRICH, HÖLLER, and LIEVENS, 2010). This setup was successfully used to predict sound pressure levels radiated by a refrigerator based on measured blocked forces of the compressor on a force-bench including proper conditioning of the suction and discharge gas pressures and temperatures.

5.4.2. Matrix Notation for TPA Methods

The transmission matrix notation is used for the deduction with the transmission matrix $\mathbf{T}$ and its entries $\mathbf{A}$, $\mathbf{B}$, $\mathbf{C}$ and $\mathbf{D}$ which are also matrices:

$$\mathbf{T} = \begin{pmatrix} \mathbf{A} & \mathbf{B} \\ \mathbf{C} & \mathbf{D} \end{pmatrix}. \tag{5.33}$$

The matrix is split into its entries in order to connect the formulation with the notation of the TPA methods. The isolator is virtually integrated into a new sound source with a different source mobility and a different blocked force and free velocity. The force and velocity at the input and output are

$$\mathbf{F}_{\text{out}} = \mathbf{A}\mathbf{F}_{\text{in}} + \mathbf{B}\mathbf{v}_{\text{in}} \tag{5.34}$$

$$\mathbf{v}_{\text{out}} = \mathbf{C}\mathbf{F}_{\text{in}} + \mathbf{D}\mathbf{v}_{\text{in}}. \tag{5.35}$$

Blocked Two-Port

The input mobility of a two port blocked at its output can be described using $\mathbf{v}_{\text{out}} = 0$ with Eq. 5.34 and Eq. 5.35 as

$$\mathbf{Y}_{\text{in,b}} = -\mathbf{D}^{-1}\mathbf{C} \tag{5.36}$$

and the blocked force at the output follows as

$$\mathbf{F}_{\text{out,b}} = \mathbf{A}\mathbf{F}_{\text{in}} + \mathbf{B}\mathbf{v}_{\text{in}} = \mathbf{A}\mathbf{F}_{\text{in}} + \mathbf{B}\left(-\mathbf{D}^{-1}\mathbf{C}\right)\mathbf{F}_{\text{in}} = \left(\mathbf{A} - \mathbf{B}\mathbf{D}^{-1}\mathbf{C}\right)\mathbf{F}_{\text{in}}. \tag{5.37}$$

The force at the input $\mathbf{F}_{\text{in}}$ can be obtained by the coupling between source and receiver as already discussed

$$\mathbf{F}_{\text{in}} = \mathbf{F}_{\text{c}} = \left(\mathbf{Y}_{\text{s}} + \mathbf{Y}_{\text{r}}\right)^{-1}\mathbf{Y}_{\text{s}}\mathbf{F}_{\text{b}} = \left(\mathbf{Y}_{\text{s}} - \mathbf{D}^{-1}\mathbf{C}\right)^{-1}\mathbf{Y}_{\text{s}}\mathbf{F}_{\text{b}}. \tag{5.38}$$

Finally, the blocked force of the source at the output of the connected isolator $\mathbf{F}_{\text{b,iso}}$ is obtained by using Eq. 5.38 in Eq. 5.37 as

$$\mathbf{F}_{\text{b,iso}} = \mathbf{F}_{\text{out,b}} = \left(\mathbf{A} - \mathbf{B}\mathbf{D}^{-1}\mathbf{C}\right)\left(\mathbf{Y}_{\text{s}} - \mathbf{D}^{-1}\mathbf{C}\right)^{-1}\mathbf{Y}_{\text{s}}\mathbf{F}_{\text{b}}. \tag{5.39}$$

Free Two-Port

The free two-port is considered in a similar manner but by using the impedance instead of the mobility to avoid singularities. An isolator approximating a rigid connection has also to approximate the result obtained without using the isolator. This can only be achieved if the matrices approximate zero and if they are connected by addition. Based on the constraint for the free two-port $\mathbf{F}_{\text{out}} = 0$ and by solving for the impedance it yields

$$\mathbf{Z}_{\text{in,f}} = -\mathbf{A}^{-1}\mathbf{B}. \tag{5.40}$$

The relation of the velocity at the input and the output of the isolator reads as

$$\mathbf{v}_{\text{out}} = (-\mathbf{C}\mathbf{A}^{-1}\mathbf{B} + \mathbf{D})\mathbf{v}_{\text{in}}. \tag{5.41}$$

Due to the coupling between the source and the receiver the velocity at the input follows as

$$\mathbf{v}_{\text{in}} = \mathbf{v}_c = \mathbf{Y}_r(\mathbf{Y}_s + \mathbf{Y}_r)^{-1}\mathbf{v}_f = (\mathbf{Z}_s + \mathbf{Z}_r)^{-1}\mathbf{Z}_s\mathbf{v}_f \tag{5.42}$$

and by inserting Eq. 5.42 in Eq. 5.41 and using $\mathbf{Y}_r = \mathbf{Z}_{\text{in},f}^{-1}$ the free velocity of the source at the output of its connected vibration isolator read as

$$\mathbf{v}_{f,\text{iso}} = (\mathbf{D} - \mathbf{C}\mathbf{A}^{-1}\mathbf{B})(-\mathbf{Y}_s\mathbf{A}^{-1}\mathbf{B} + 1)^{-1}\mathbf{v}_f. \tag{5.43}$$

Transformed Source with Vibration Isolator

The source mobility is the relation between the blocked force and the free velocity as

$$\mathbf{v}_{f,\text{iso}} = \mathbf{Y}_{s,\text{iso}}\mathbf{F}_{b,\text{iso}}. \tag{5.44}$$

By using Eq. 5.39 and Eq. 5.43 the source mobility of the new source including the vibration isolator is calculated as

$$(\mathbf{D} - \mathbf{C}\mathbf{A}^{-1}\mathbf{B})(-\mathbf{Y}_s\mathbf{A}^{-1}\mathbf{B} + 1)^{-1}\mathbf{v}_f = \mathbf{Y}_{s,\text{iso}}\left(\mathbf{A} - \mathbf{B}\mathbf{D}^{-1}\mathbf{C}\right)(\mathbf{Y}_s - \mathbf{D}^{-1}\mathbf{C})^{-1}\mathbf{Y}_s\mathbf{F}_b \tag{5.45}$$

where one inversion can be avoided by summarizing two terms:

$$\mathbf{Y}_{s,\text{iso}} = (\mathbf{D} - \mathbf{C}\mathbf{A}^{-1}\mathbf{B})(-\mathbf{Y}_s\mathbf{A}^{-1}\mathbf{B} + 1)^{-1}(\mathbf{Y}_s - \mathbf{D}^{-1}\mathbf{C})(\mathbf{A} - \mathbf{B}\mathbf{D}^{-1}\mathbf{C})^{-1}. \tag{5.46}$$

The integration of the vibration isolators into a new source with arbitrary degrees of freedom, connection points and number is isolators is fully described by Eq. 5.39 transforming the blocked force or by Eq. 5.43 transforming the free velocity of the source and the transformation of the source mobility in Eq. 5.46. These matrix equation capture the multi-dimensional case similar to the one-dimensional case described in Eq. 5.30.

5.5. Application I—Uncertainty in Sensor Position

It was shown that the results obtained by the TPA methods could be theoretically transferred into each other in Section 5.3.3. This mainly involves multiplication with mobility matrices or their inverse. In order to investigate the problems introduced by this calculation the problem is simplified to only one mobility matrix. The measurement of a transfer path with either one of the discussed TPA methods inherently involves both mobility matrices of the source and the receiver. For the transform of the results it is assumed that the mobility matrix multiplied by its inverse equals the identity matrix $\mathbf{I}$. This assumption is only valid if the measured system is time-invariant and if the measurement positions remain identical[5]. Only the latter aspect is investigated, as changes in the position while placing the sensors multiple times are practically unavoidable. This can be characterized by a position uncertainty similar to the example in Section 4.4.3.

As an exemplary structure-borne scenario an analytic model for the driving-point and transfer mobilities of a thin rectangular plate with geometry $0.8\,\text{m} \times 0.5\,\text{m}$ corresponding to the ceiling of the airborne-scenario L_1 in Section 4.3 was used. The simply supported boundary condition (pinned-pinned) is chosen resulting in sinusoidal mode shapes with vanishing normal velocity at the boundary. This model is similar as the one used for the airborne scenario and described in detail in (F. FAHY and WALKER, 2004) with a modal superposition approach in the section *out-of-plane vibrations of rectangular plates*[6]. For the chosen example the degrees of freedom can be reduced as only the force in normal direction with two moments and the responding velocity in normal direction and two angular velocities have to be considered. The material properties are chosen for an aluminum plate with Young's modulus of $69 \cdot 10^9\,\text{N/m}^2$, Poisson ratio of 0.346, a density of $2710\,\text{kg/m}^3$ and a frequency independent loss factor of $0.1\,\%$.

The mobility matrix for a single point with the coordinates $(0.27\,\text{m}, 0.17\,\text{m})$ is shown in Figure 5.6. The driving point mobilities shown on the diagonal are all minimum-phase and the phase is limited to $\pm 90°$ according to the theory (F. FAHY and GARDONIO, 2007) where the remaining components are not minimum-phase. The coupling between the degrees of freedom reaches its minimum for coordinates in the center of the plate and increases towards the sides. For the chosen position the coupling is in the same order of magnitude as the diagonal components. The rotational mobilities have less resonances than the translational mobilities. This is explained by the derivation used to apply moments at one point similar to dipole excitation in the airborne scenario. Once again the Common Acoustic Poles and Zeros approach (CAPZ) approach is valid as the poles being the modes are common for all degrees of freedom and all contact points of a

---

[5]Measurement noise and high condition numbers can additionally become problematic (THITE and THOMPSON, 2003).

[6]Since no modifications are applied to this model the equations are not repeated in this thesis

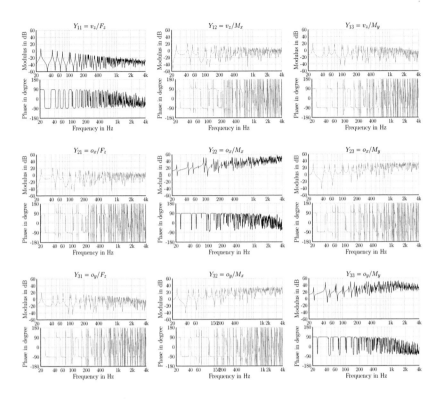

Figure 5.6.: Mobility matrix of a rectangular plate with geometry $0.8\,\mathrm{m} \times 0.5\,\mathrm{m} \times 1\,\mathrm{mm}$. The spectra on the diagonal (black) are the driving point mobilities whereas the remaining spectra (gray) describe the interaction between the degrees of freedom. The magnitude is calculated with $20\log_{10}(|Y|)$ in dB re $\mathrm{m/Ns}$, $1/\mathrm{Nms}$ and $1/\mathrm{Ns}$, respectively.

structure. In case of the rotational components specific modal coefficients are always zero but this does not conflict with the CAPZ approach.

The position was exemplarily altered by $0.1\,\mathrm{mm}$, $1\,\mathrm{mm}$ and $1\,\mathrm{cm}$ towards the center of the plate. The spectral deviation of the translational driving point mobility is depicted in Figure 5.7. With increasing uncertainty in position the deviation in the mobility increases as well, where the resonances can also be observed in this deviation similar to the room acoustic example. It is important to mention, that the error in contrast to the airborne example does not increase drastically over frequency. This is explained by the lower increase of the modal density over frequency and the different dependence of the modal coefficients on

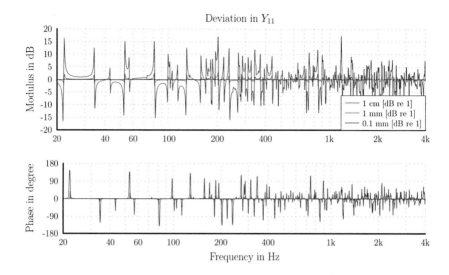

Figure 5.7.: Deviation obtained by spectral division of translational driving point mobility on a thin aluminum plate for a shift in position by 0.1 mm, 1 mm and 1 cm for one specific position on the plate.

both frequency and position in the plate model. MC simulations with 50 run for positions distributed over the entire plates and the same perturbations in position were used to obtain the mean error over frequency as depicted in Figure 5.8. This mean error is smaller than the peaks observed in the example for a particular position. Moreover, this provides already an approximate measure on the errors to be expected. This error does not show a strong dependence on frequency as mentioned before, but it has to be pointed out, that the results are only valid for the specific example studied. In particular, the uncertainty in the mobility is dependent on the geometry and the damping as already investigated for the airborne scenario.

The position uncertainty observed in practical measurement setup is assumed to be in the range of a few millimeters if measurements are conducted with an impedance hammer. Hence, mean errors in the translational driving point mobility of approx. $\pm 0.5\,\text{dB}$ have to be expected. Since this is not the only source of uncertainty the combined uncertainty has to be expected to be in the range of a few dB. By comparing these uncertainties with the results obtained in (LIEVENS, 2013) for detailed measurements of a washing machine on a wooden floor and further estimating a position uncertainty of $3 - 5\,\text{mm}$ between mobility

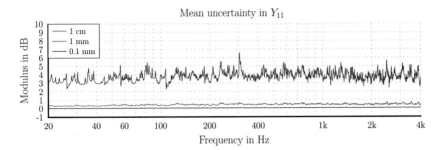

Figure 5.8.: Mean error in translational driving point mobility for thin aluminum plate obtained by MC simulations with 50 runs for a position uncertainty of 0.1 mm, 1 mm and 1 cm.

and in-situ measurement the order of magnitude seems to be reasonably in the same order of magnitude although the material properties and geometries are different[7].

5.6. Application II—Modeling of Typical Simplifications

To decrease the measurement complexity simplifications are applied in practice. (LIEVENS, 2013) investigated uncertainties caused by assumption about the source based on 1-D measurements in detail. These simplifications can also be modeled by modifying the matrices introduced above where these modifications are exemplarily applied to Eq. 5.15. This approach is only applicable if the full matrices and the input signals are known. Hence, this is mainly interesting from a theoretical point of view or by using numerical simulations instead of extensive measurements that also include measurement noise and the uncertainty in positioning of the sensors and actuators as discussed before. As a measurement of the coupled system in operating condition generally involves all components the errors can therefore be studied by comparing the result obtained by using no simplifications to the ones obtained with the modeled simplifications.

As an example, the analytic model of the thin rectangular plate is used for both source and receiver mobility as also described in the appendix A.7. The geometry of the source plate was chosen as 0.4 m × 0.3 m with a thickness of 5 mm whereas the geometry of the receiver plate and the remaining parameters

---

[7]Experiments with an airborne enclosure of geometry L_1 and an aluminum plate with the same properties as the plate studied showed similar results but are not listed here further.

were the same as in the previous example. Both plates are connected by two contact points spaced by a small distance of 4 mm near the center of both plates as illustrated in Figure 5.9 to capture a moderate coupling scenario. Moreover, the two translational source and receiver driving point mobilities are very similar due to the small distance of the contact points on the plate.

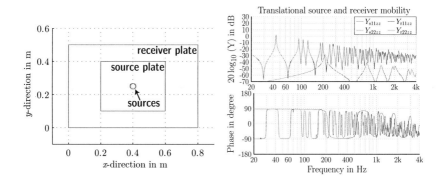

Figure 5.9.: Simplified setup (left) for source (blue) and receiver (red) represented by rectangular plates where the two green circles represent the coupling points. Translational driving point mobility for contact point 1 and 2 for the source and the receiver.

Degrees of Freedom Neglecting degrees of freedom can be modeled by deleting entries (rows) in the blocked force vector $\mathbf{F}_b$ and be deleting rows and columns in the mobility matrices and hence in the coupling matrix $\mathbf{C} = (\mathbf{Y}_s + \mathbf{Y}_r)^{-1} \cdot \mathbf{Y}_s$. Two different errors are introduced. Firstly, the introduced force causing a path contribution at the receiving point is missing in the synthesis. Secondly, the interaction between the neglected degrees of freedom with the remaining degrees of freedom changes the coupling matrix. Hence, path contributions of the remaining degrees of freedom could be overestimated or underestimated. As an example, a source that only produces a force in normal direction and that has additionally a mobility matrix $\mathbf{Y}_s$ that is diagonal cannot be modeled correctly by neglecting all degrees of freedom except for the normal translational component. This simplification reduces the measurement complexity significantly as less or simpler sensors have to be positioned and less degrees of freedom have to be excited via impact hammer or shaker yielding a shorter overall measurement duration.

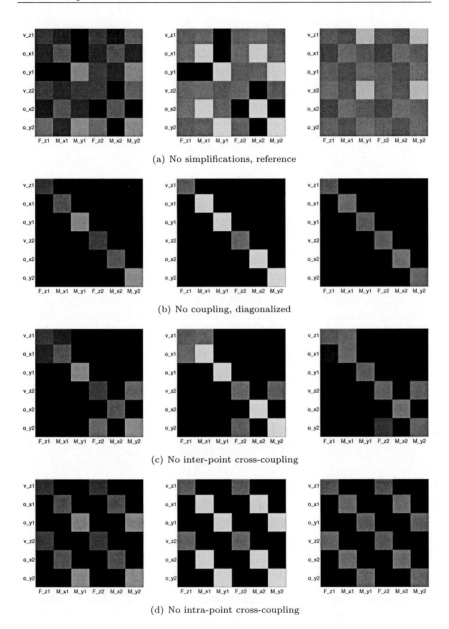

(a) No simplifications, reference

(b) No coupling, diagonalized

(c) No inter-point cross-coupling

(d) No intra-point cross-coupling

Figure 5.10.: Schematic plot of source (left, $\mathbf{Y}_s$) and receiver (center, $\mathbf{Y}_r$) mobility matrices and resulting coupling matrix (right, $\mathbf{C}$) of the basic example using rectangular plates and two connection points. The color represents the amplitude of the entries in dB (dark: low, bright: high values).

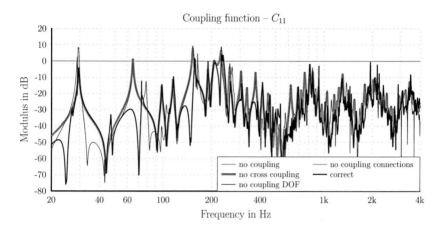

Figure 5.11.: Transversal coupling function for connection point 1 ($\mathbf{C}_{11}$) with typical simplifications applied to the source and receiver mobility matrices.

Without knowledge of the source signals only the interaction error can be analyzed. This is similar to the error obtained for neglecting the intra-connection cross coupling but only the elements beside the diagonal of the corresponding degrees of freedom are set to zero. This remains the full size of the matrix.

No Source-Receiver Coupling If the source is considered as an ideal force source, the interaction between source and receiver is neglected. Hence, the coupling matrix simplifies to the identity matrix $\mathbf{C} = \mathbf{I}$.

No Inter-connection Cross-coupling Neglecting the coupling between connection points can be modeled by using the sub-matrices from Eq. 5.2 for the source and the receiver. The coupling between the degrees of freedom at one contact point is still considered and hence the sub-matrices $\mathbf{Y}_{\mathrm{el},ii}$ are kept but the cross sub-matrices are overwritten: $\mathbf{Y}_{\mathrm{el},ij} = 0$.

No Intra-connection Cross-coupling Keeping the cross-coupling between the contact points and neglecting the coupling between the degrees of freedom is equivalent to using only the diagonal of all sub-matrices: $\mathrm{diag}\left(\mathbf{Y}_{\mathrm{el},ii}\right)$ and $\mathrm{diag}\left(\mathbf{Y}_{\mathrm{el},ij}\right)$.

No Cross-coupling By neglecting all cross-coupling between the contact points and between the degrees of freedom the mobility matrices are replaced by their diagonal entries: $\mathrm{diag}\,(\mathbf{Y_s})$ and $\mathrm{diag}\,(\mathbf{Y_r})$. As a consequence, the corresponding impedance matrices are also diagonal and all entries are minimum-phase, as the mobilities on the diagonal are all minimum-phase. Furthermore, the resulting coupling matrix is diagonal and minimum-phase and the entries could be calculated by using only the corresponding entries from the source and receiver mobility matrix.

The principal symmetry of the matrices with applied simplifications is exemplarily simulated in Figure 5.10 for a frequency of 400 Hz. The magnitude is represented by the brightness of the color but has to be understood an illustrative and qualitative manner only. A general rule for the matrices can be given based on these examples. If both the source and the receiver mobility matrix have the same symmetry this symmetry also follows for the coupling matrix. Furthermore, a comparison of the coupling function in terms of the spectrum of the translational component of connection one $(\mathbf{C_{11}})$ is depicted in Figure 5.11. As can be seen, the results vary over frequency and the assumption of a negligible source receiver coupling is not valid. Moreover, the result without coupling at all and by neglecting only the coupling between the degrees of freedom are very similar for this particular example.

Although this model is basic and free of noise the deviations are in the order of $10 - 30\,\mathrm{dB}$ for some frequencies. In practice, the energetic error, e.g., averaged over an octave band would be smaller. This is a valid approach for broadband input signals. If multi-tone signals as, e.g., generated by electrical machines are found at the input, the errors are higher and can be in the order of magnitude as presented. As this is only one example it should be understood as a possible guideline to model the influences only and the results from this example should not be directly generalized.

At this point the link to the CAPZ is important. There is one set of poles for all entries of the source mobility and one set for the receiver mobilities. The connection of the two structures yields modified resonance frequencies. Moreover, the zeros of the entries have an influence on the position of the poles in the matrix calculations. While the coupling function for the diagonalized case is still very smooth it is more complicated for the calculation using full matrices due to the influence of the zeros of the components beside the diagonal of both mobility matrices.

5.7. Summary and Scientific Contribution

A theoretic approach to investigate the uncertainties in the measurement and modeling of structure-borne sound propagation has been presented where the interaction between structure-borne sound sources and connected systems was modeled using mobility matrices. The principal symmetries occurring in these matrices have been discussed as well as the relation to the CAPZ approach. Since mobilities and impedances are mainly analyzed in the frequency domain the phase constraints and the causality was studied. Driving point mobilities and impedances of linear and passive systems are found to be minimum-phase with a phase in the range of $\pm 90°$ and hence both causal. Cross and transfer mobilities are causal whereas the impedances are generally not causal.

A matrix notation for three different TPA methods—classic TPA, measurement of the coupled system and operational TPA—has been deduced. The obtained transfer paths by the method have been found to be not identical and hence the theoretic differences have been discussed. Moreover, relations between the transfer paths and a transformation of these result have been presented involving additional measurements of mobility matrices. Furthermore, the two-port approach for vibration isolators has been integrated for an arbitrary number of contact points.

Based on this notation an application example using an analytic model for thin rectangular plates has been investigated. The uncertainty in the mobility matrix due to uncertainty in positioning the sensors was simulated by means of Monte Carlo simulations and results have been compared to the airborne scenario. As a main difference the uncertainty was found to be mainly independent on frequency for the specific example chosen.

Since simplifications to the general 6 degree of freedom case with several contact points are commonly applied in practice to decrease the measurement complexity the influence on the mobility matrices of sources and the receiver have been investigated. The deviation in the source receiver coupling was presented for one example. It can be stated that neglecting degrees of freedom or cross-coupling can have severe effects on the prediction of the structure-borne sound propagation if the entries and especially the cross-coupling entries in the mobility matrix are in the same order of magnitude as the main driving point mobilities.

In summary, large errors in the order of a few dB have to expected for reasonable uncertainty in positioning of sensors of approx. $1 - 5\,\text{mm}$ and applying typical simplifications to reduce the measurement effort, e.g., by neglecting cross-coupling.

In general, a similar behavior of the uncertainties in the transfer function caused by inaccuracies in the sensor placement as for the airborne application is assumed. However, more research using different structure models is required to proof this.

6

Conclusion and Outlook

Acoustic transfer functions and their measurement were investigated in a detailed and theoretical manner. Therefore, basics in signal processing, representations of transfer functions and the fundamentals in uncertainty modeling were presented. Afterwards, the uncertainties have been exemplarily studied and divided into artifacts caused by the measurement method and the measurement equipment with a special focus on signal processing and perturbations caused by misplacement of observation points and further non ideal characteristics of the actuators.

It can be stated that the measurement of acoustic transfer functions remains a challenging task although the basic theory has been applied for decades. Mainly the measurement equipment in terms of a limited linear range of operation causes artifacts that are not fully discovered and sometimes claimed to be negligible. A nonlinear modeling technique was exploited in order to analyze the principal influences on impulse responses measurement with exponential sweeps. Although often claimed else, nonlinearities cannot be fully suppressed by applying a simple time window to the measured impulse response including distortion. Moreover, a further artifact additional to the usually observed harmonics in terms of inter-modulation distortion in sweep measurements has been found. Although the nonlinear modeling is widely used to emulate, e.g., for guitar tube amplifiers driven in a highly nonlinear range with acceptable audio quality, the post-processing of measured distorted impulse responses to obtain the linear impulse response was found to be unstable and hence not applicable to reduce the observed artifacts with sufficient accuracy. Moreover, the influences of these distortion artifacts on a simulated room impulse response and especially the derived room acoustic parameters have shown that the nonlinearities need further investigation and might lead to uncertainties in the parameters in the order of the just noticeable difference. For this purpose an entire measurement chain for typical acoustic measurements was implemented, that is capable of emulating the artifacts caused by noise, quantization or the nonlinearities correctly. All parameters in this open-source MATLAB model can be controlled in terms of single number values and form hence a tool also for future analysis of uncertainty factors. The underlying ideal linear impulse response of the device under test can hereby be captured by either measurement or simulation data, e.g., from analytical models of structure-borne or airborne transfer functions as applied already in this thesis.

Airborne transfer paths have been successfully modeled by means of combination and adaptation of existing analytic transfer function models. For the ongoing uncertainty analysis the contributions caused by the limited accuracy in the positioning of sensors and the orientation of directive sound sources has been exemplarily studied. With the advantage to provide the possibility of controlling all parameters independently in the analytical models with low computation times this approach outperforms extensive and time consuming round robin tests. It has been shown for an application example to measure the in-situ reflection index that these simulations can be further used to study the robustness of the evaluation procedure against unavoidable perturbations in the field and derive constraints for the hardware, e.g., in terms of claiming a certain maximum directivity, to ensure a certain range of uncertainty. Moreover, systematic errors can be detected by comparing the results of the model with the ideal outcome without perturbations. Afterwards, round robin measurements could provide a practical measure for the combined uncertainty in the field.

An analytical model for rectangular rooms has been studied in detail capable of delivering accurate results comparable to validated finite element simulations if certain requirements are met. Furthermore, these models are advantageous as they are suitable to characterize the uncertainty in the transfer function itself. The relation between the modal parameters and the pole-zero model from the field of signal processing providing a compact description of transfer functions has been shown. In order to investigate if this basic and theoretic modeling approach is compatible with practical measurements, a comparison with measurement data has been presented. Additionally, the link transferring measurement data into this compact description hence being suitable for further uncertainty investigation has been investigated in terms of a rational fit algorithm. Moreover, the derivation of room acoustic parameters from impulse responses with this analytic model and further perturbations in the position of the sensors has been shown to match with real-life measurements of a scanned area in terms of uncertainties. By adapting this analytic model to additionally capture directivities of sources and receivers by means of physical multipoles, the influence of the orientation accuracy on both the modal parameters and room acoustic parameters has been studied for basic directivity patterns. As the studied model employs the modal superposition approach the advantage in terms of fast computation times compared to other simulation methods is found in the lower frequency range only where the modal density and especially the number of modes is fairly low.

As a key result in this context the common acoustic pole and zero approach is applicable leading to the statement that changes in the position of sensors and sound sources, their directivity and also changes in the orientation of the directive patters do not effect the eigenfrequencies nor the modal damping but only the zeros observed in the transfer path and in other words, the weighting of the modes in the superposition only. Changes in geometry or, e.g., temperature have an influence on the eigenfrequencies and potentially on the damping. Again, this approach, i.e. the fitting of measurement data and the relation between modal parameters and the poles and zero model, can only be applied with reasonable computation times and accuracy for a limited number of modes and hence up to an upper frequency bound.

A detailed investigation of uncertainty contributions for the measurement and modeling of structure-borne sound propagation has been found in the literature. The approach used for the airborne scenario has been transferred to a structure-borne example using an analytical model for out-of-plane vibrations of thin rectangular plates to provide a possible link between both fields of application. In an application example the uncertainty in the position of sensors for structure-borne mobility measurements has been investigated by means of Monte Carlo simulations. As the results indicate high uncertainties in the mobilities the accuracy in positioning sensors is found to be critical besides the remaining uncertainty contributions found in literature regarding time-variances or aging of structure-borne systems and deviations between measured specimen. It can be stated that the uncertainty in the prediction of sound pressure levels by a structure-borne scenario compared to an airborne scenario is reasonable higher by keeping similar accuracy in positioning of sensors additionally caused by unavoidable negligence of degrees of freedoms due to practical measurement and modeling reasons. In this context, the coupling between the sound source and the structure has a in general an influence on the sound propagation. The cross-coupling between degrees of freedom and multiple connections points between the source and the receiver and especially the influence due to negligence of this cross-coupling has been investigated leading to large uncertainties. Although a theoretic relation between commonly used methods for transfer path analysis and synthesis has been deduced in terms of a matrix notation a conversion of the results obtained by the different methods into each other is not considered to deliver reliable results in practice. This is explained by potentially ill-conditioned inversion problems and limited signal to noise ratios in practice. Nevertheless, the implementation of such a basic structure-borne scenario is considered a powerful tool as, e.g., also used for novel structure borne sound power measurements, to investigate the principal uncertainties in structure-borne sound modeling and to study the principal outcome and the robustness of novel measurement and modeling methods as, e.g., the operational transfer path analysis.

The applicability of the common acoustic pole and zero approach has been shown for the basic example of the rectangular plate. However, due to the matrix inversion required to model the interaction between the source and the receiver including cross-coupling, the zeros of mobilities interact with the poles. This is reasonable since applying a load—this is generally dependent on the position on the connection point on the structure—to a structure changes it eigenfrequencies and hence the poles. It remains to be solved how this interaction between the poles and zeros and therefore also the position accuracy influencing the zeros propagates through this matrix inversion. Finally this could lead to an uncertainty in both poles and zeros of this coupling transfer function.

Although a tool chain using an emulation of an acoustic measurement chain and the implementation of analytic models with single number input parameters has been developed the potential of such a modeling approach and especially the assessment of the combined measurement uncertainty has not been fully exploited. The studied examples are only an excerpt of possible applications. The calculation of the uncertainty budget via this modeling approach for at least the low frequency range in room acoustics and moreover the combination with, e.g., the mirror source method and further statistical models to predict the uncertainty due to the directivity of real-life measurement loudspeakers has to be solved. Furthermore, the developed measurement method for room impulse responses with arbitrary radiation patterns has to be extended towards higher frequencies and higher spherical harmonic orders and compared to the adapted analytic model for rectangular rooms, the image source method including directivity and finally measurements. More experience in dealing with acoustic transfer functions in terms of a pole-zero approach including uncertainty in the position of these poles and zeros and the propagation through evaluation methods onto the derived parameters has to be gained. All in all, only a small step towards the challenging and interesting field of the assessment of uncertainties in acoustics has been made so far.

With the findings provided in this thesis, more measurement scenarios both airborne and structure-borne should be analyzed to yield general statements of the uncertainties of the transfer functions. Besides the positioning and orientation uncertainty studied, more input quantities should be investigated in detail that are known to have an influence, e.g., temperature, variation in geometry or variation in the condition of the device under test. Especially in the field of structure-borne sound, e.g., in automotive industry, it is known that variation in production leads to uncertainties in the transfer functions for a specific product. These uncertainties could be also integrated in the simulation scenario.

For specific signal number parameters, e.g., sound pressure levels or room acoustic parameters, the just noticeable differences have been measured providing a measure to relate the uncertainties to the perception. Although the relation between complex transfer functions and in particular their uncertainty to the perception is far more complicated, research in this direction is required to not only state that a measurement is sufficiently precise from a technical but also from a perceptional point of view.

Acknowledgements

I would like to express man gratitude to those who supported me up to this point and finally made this work possible.

First of all I like to thank Johanna Holsten for giving me space and time to work overhours and additionally giving me motivation and valuable feedback throughout the years. Thanks goes of course to my loving family for the great support. I would like to thank Prof. Michael Vorländer for the possiblity to work freely on this thesis and also many other interesting and challenging projects. Thanks go also to Prof. Kay Hameyer for taking the role as co-referee and Dr. Gottfried Behler for many hours of valuable discussion regarding all engineering topics.

Special thanks go to my colleagues at ITA. Especially to the amazing spirit of our ITA-Toolbox developer team: Markus Müller-Trapet, Dr. Roman Scharrer, Dr. Bruno Masiero, Martin Guski, Johannes Klein and especially Martin Pollow for his spherical know-how. Thanks to Frank Wefers for fast, bilateral communication with low latency about signal processing. I would like to express my gratitude to Ingo Witew, Renzo Vitale, Dr. Marc Aretz, Timo Lohmann and Christoph Höller for valuable discussions and especially Ramona Bomhardt who is completely fearless of ugly equations. Thanks to Dr. Matthias Lievens who taught me to be suspicious of any measurement result and who inspired me by knowing so many details of so many publications. Furthermore, I am thankful for the work of all my students—especially Julian Blum, Benedikt Krechel and Martin Kunkemöller. Thanks to Rolf Kaldenbach for this constructive cooperation in all kinds of electronics' design.

I would like to further express my gratitude to my friends and colleagues outside ITA for great discussions and feedback: Lothar Kurz, Dr. Bernd Tessendorf, David Franck, Dr. Michael van der Giet, Prof. Stephan Paul, William Fonseca, Edmar Baars and Prof. Arcanjo Lenzi.

A

Appendix

A.1. The ITA-Toolbox for MATLAB

The ITA-Toolbox is a set of functions, classes and scripts for MATLAB to manage tasks in the field of acoustics. It can be used on any operating system which is supported by MATLAB. By the time of the submission of this thesis it was compatible with all version from R2010a to R2013b and it is expected that this toolbox is further developed. Main parts of this toolbox are open-source but some specific optional packages, e.g., for spherical harmonics calculations, as used in thesis have not been released to the public. However, most routines, simulations scripts and plotting scripts developed during this thesis are included in the free version.

A.1.1. Short History

In 2007 a basic set of MATLAB routines was implemented based on a simple data structure. This structure was mainly influence by the file structure used in the loudspeaker measurement program *Monkey Forest (MF)*. The functionality was simple but efficient by transforming objects from time to frequency domain and vice versa and plot the data in these domains with different properties. A small set of meta data was already stored inside these objects containing sampling rate, comments and some plot settings. This package called *Monkey MATLAB* was already used in (DIETRICH, 2007) to import and export data from MF, plot, apply pre-processed band-pass filters and finally calculate room acoustic parameters. Function calls were as simple as possible at that time and tried to guess options not specified by the user.

During an industry project in 2008 this package was improved in both functionality and the data structure used for the objects to allow more flexible import and export to other formats and basic calculations for transfer path analysis and synthesis. In 2009 this side project *Monkey MATLAB* was improved by routines already implemented in (POLLOW, 2007). Finally, this side project gained more popularity inside the Institute of Technical Acoustics at RWTH Aachen University and the LVA at the Federal University in Santa Catarina. This package was later renamed to *ITA-Toolbox* as the focus changed from the original idea. The unified prefix used for all implemented MATLAB files was changed to `ita_`. With the use of a revision control system (CVS and later SVN), a bug tracking system (TRAC) and a wider range of frequent users inside the Institute of Technical Acoustics the stability was increased quickly. Although this toolbox was primarily not intended for such a large number of users, it became the essential component for almost all scientific and industrial projects at the Institute of Technical Acoustics dealing with signal processing, data plotting and measurements.

In 2011 an open-source kernel of the ITA-Toolbox was released on *www.ita-toolbox.org* with a BSD license. In 2012 further applications for measurements, room acoustic analysis and nonlinear modeling were added.

A.1.2. Functionality and Concept

The core of the ITA-Toolbox consists of the classes `itaValue`, `itaResult` and `itaAudio`[1] Acoustic measurement or simulation data is commonly stored numerically in vectors or matrices. Associated with this data is meta information, e.g., sampling rates, comments, coordinates, domain or physical units. Programming a container for all these different types of data was realized efficiently by using the concept of Object-Oriented Programming (OOP) in MATLAB.

Based on the need to calculate with physical symbols including their physical unit the class `itaValue` has been developed. The objects of the class store the value along with a physical unit. It is capable of calculating values and units for multiplication and division of two variables.

Coordinates are indeed a good example to illustrate the benefit of using OOP for transformation purposes, e.g., between Cartesian to cylindrical or spherical coordinates. The object stores the data along with the information of the domain

---

[1] Parts of the description has been published in (DIETRICH et al., 2013; DIETRICH et al., 2012; DIETRICH, GUSKI, and VORLÄNDER, 2013)

of representation with the class (`itaCoordinates`). By simply using `.cart` for Cartesian or `.sph` for spherical representation the data is converted according to its current and the target representation.

Audio data is commonly much more complicated than single values or coordinates. The same concept is therefore used and extended towards a class called `itaAudio` that stores equidistantly sampled audio data from measurements or simulations in either time or frequency domain where the time domain data can always be accessed by `.time` and the frequency data by `.freq`. The according time stamps or the frequency vector for plots are calculated by `.timeVector` and `.freqVector` respectively. Furthermore, simple mathematical operations, e.g., multiplication (`*`), division (`/`), summation and subtraction (`+,-`) are implemented for these audio objects. This enables to directly write formulations in a text book manner. Hence, multiplications and divisions are realized in the frequency domain. The basic functionality and some plots along with their function call are summarized in Figure A.1.

For post processing of measured or simulated acoustic data several standard routines are available, e.g., time-windowing, bandpass or fractional octave band filtering, cross-blending or scaling. The advantage of using ITA-Toolbox routines lies mainly in the meta data stored along with the audio data and the simple function calls. There is no need to specify the sampling rate when filtering the data, as it is automatically read from the object itself. All functions, methods and properties of the classes can be accessed via the command line of MATLAB, with direct access to data in both time and frequency domain always possible without explicit Fourier transform.

A.1.3. Professional Sound Boards—Hardware Communication

The communication with professional sound cards on Windows (ASIO, Steinberg ASIO SDK), Linux (ALSA) and Mac OS X (CoreAudio) is realized using open-source *PortAudio* (*www.portaudio.com*) and the bridge to MATLAB via *Playrec* (*www.playrec.co.uk*). *These names are trademarks even if not marked as such.* The required source code, the software license texts and an adapted compilation script as well as pre-compiled `MEX`-files are included in the open-source version. For the measurements in this thesis the following sound cards were successfully used (without rating and without comments on stability issues):

- *Presonus FireBox* (Firewire);

- *Presonus LightPipe* (Firewire);
- *Presonus FP10* (Firewire);
- *RME Multiface* (HDSP);
- *RME Digiface* (HDSP).

Specific audio measurement hardware including remote controllable microphone pre-amplifiers and power amplifiers developed by the Institute of Technical Acoustics at RWTH Aachen University (ITA) was used. Two devices developed by the ITA called *ROBO* (only analog signal conditioning) and *ModulITA* (including an *RME Multiface*) and a device developed by Swen Müller, Brazil called *Aurelio* have been used. Remote control was implemented in MATLAB and the communication was realized via *portMidi* (*portmedia.sourceforge.net/portmidi/*). These routines are open-source. Additional measurement hardware only available at the ITA, e.g., controllable turntables and a rotating arm for directivity measurements are connected via *RS232* and *RS485*. Hence, these routines are not released to the public.

A.1.4. Measurement Classes

The application *Measurement* features several classes for specific measurement setups (MS). `itaMSRecord` is used to simple record sound directly into an `itaAudio` on specific sound card channels with the sampling rate and the length specified inside the object. The inherited class `itaMSPlaybackRecord` adds functionality for simultaneous multi-channel playback while recording. The class `itaMSTF` (Measurement Setup Transfer Function) additionally realizes correlation measurements, e.g., involving deconvolution techniques. These measurement objects store the excitation signal that can be freely defined by the user, but especially sweeps and MLS are used. A so-called *compensation* which is the inversion of the complex spectrum of this excitation signal is generated automatically in the background. The level of the measurement can be controlled in this object by setting the *output amplification.* This value is automatically accounted for by reciprocally scaling the measured input signal. Hence, the absolute values of the measured impulse response remain constant regardless off the actual output amplification of the measurement. Only the signal-to-noise-ratio (SNR) will change accordingly.

Figure A.2 shows basic GUI elements of the measurement application and the MATLAB command line window with a summary of the measurement object.

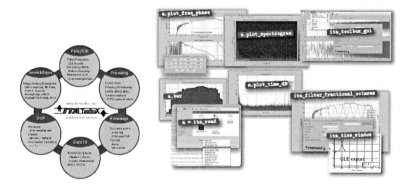

Figure A.1.: Overview of the basic functionality of the ITA-Toolbox (left) and illustrative plot examples with object-oriented function calls (right).

Parameters can be easily modified in a GUI using the function .edit. The measurement itself is triggered by the method .run and directly yields the impulse response as the deconvolution is applied to the measurement object.

The inherited class itaMSTFdummy emulates an entire measurement chain (c.f. Figure 3.16) including quantization effects, measurement noise, linear and nonlinear system characteristics as described in detail in Section 3.4 and hence not sound card is required. A tutorial script called ita_tutorial_measurement.m is available to illustrate the functionality of the measurement objects and the emulation of the measurement chain.

A.2. Rotation of Physical Multipoles

The rotation of the dipole or quadrupole around the z-axis by an angle ϕ in radiants is implemented as follows. Two auxiliary variables are used:

$$
\begin{aligned}
c_{\mathrm{a}} &= \frac{c_{(2,0,0)} + c_{(0,2,0)}}{2} \\
c_{\mathrm{b}} &= \frac{c_{(2,0,0)} - c_{(0,2,0)}}{2}
\end{aligned}
\tag{A.1}
$$

171

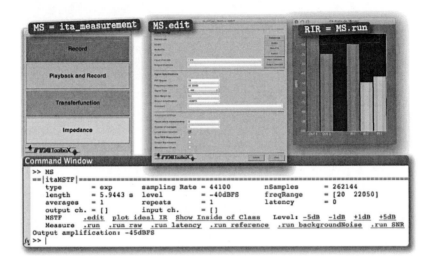

Figure A.2.: Screen-shots of GUI elements of the measurement application inside the ITA-Toolbox and MATLAB command window showing a summary of the parameters of the `itaMSTF` measurement object.

The rotation of the dipole $(k, l, m) = (1, 0, 0)$ and the quadrupoles read as

$$
\begin{aligned}
c_{(1,0,0),\text{rot}} &= \cos(\phi) & \cdot c_{(1,0,0)} + & \quad \sin\phi \cdot c_{(0,1,0)} \\
c_{(2,0,0),\text{rot}} &= \cos(2\phi) & \cdot c_b + & \quad \sin 2\phi \cdot c_{(1,1,0)} + c_a \\
c_{(1,1,0),\text{rot}} &= \cos(2\phi) & \cdot c_{(1,1,0)} - & \quad \sin 2\phi \cdot c_b \\
c_{(0,1,1),\text{rot}} &= \cos(\phi) & \cdot c_{(0,1,1)} - & \quad \sin\phi \cdot c_{(1,0,1)}
\end{aligned}
\tag{A.2}
$$

where c could be the room impulse response or the spectrum or a vector containing the S H coefficients of the multipoles.

A.3. Room Acoustic Parameters and Modal Superposition

Problems were observed in calculating room acoustic parameters based on simulation results with known modal reverberation times as the obtained reverberation times according to ISO 3382 do not correspond to modal decays. Furthermore, the obtained reverberation times are dependent on the position of the source and the receiver especially for low frequencies. The parameters clarity index and hence also definition can be obtained from this Schroeder curve as well.

The modal superposition according to Eq. 4.13 can be investigated directly in time domain for a maximum number of modes N as

$$h(t) = \sum_{i=1}^{N} h_i(t) \tag{A.3}$$

with the impulse response of a single room mode as

$$h_i(t) = \mathcal{R}\left(a_i e^{-t/\tau_i} e^{\mathrm{j}\omega_i t} e^{\mathrm{j}\phi_i} \cdot \varepsilon(t)\right) = a_i e^{-t/\tau_i} \cos(\omega_i t + \phi_i)\varepsilon(t) \tag{A.4}$$

with a_i the modal coefficient, $\tau_i = {}^{\mathrm{RT}_i}/_{3\ln(10)}$ the modal decay time expressed as the half-value period, ω_i the eigenfrequency, ϕ_i the starting phase and $\varepsilon(t)$ the *Heaviside* function.

A.3.1. The Schroeder Curve and Room Modes

The Schroeder curve or energy decay curve is defined, e.g., in (ISO 3382, 2009) using a backwards integration as

$$L(t) = \int_t^\infty h^2(t')\,\mathrm{d}t'. \tag{A.5}$$

By using Eq. A.3 we obtain

$$L(t) = \int_t^\infty \left(\sum_{i=1}^N h_i(t')\right)^2 \mathrm{d}t'. \tag{A.6}$$

and splitting the terms of the integrand into mono-frequent and mixed frequency terms yields

$$L(t) = \int_t^\infty \sum_{i=1}^N \underbrace{h_i^2(t')}_{\geq 0}\,\mathrm{d}t' + 2\int_t^\infty \sum_{i=1}^N \sum_{j=i+1}^N h_i(t') \cdot h_j(t')\,\mathrm{d}t'. \tag{A.7}$$

By commutating summation and integration this transforms to

$$L(t) = \sum_{i=1}^N \underbrace{\int_t^\infty h_i^2(t')\,\mathrm{d}t'}_{L_{ii}} + \sum_{i=1}^N \sum_{j=i+1}^N \underbrace{\int_t^\infty 2\,h_i(t') \cdot h_j(t')\,\mathrm{d}t'}_{L_{ij}} \tag{A.8}$$

with the mono-frequent terms L_{ii} and the mixed terms L_{ij}.

It is commonly expected that the Schroeder decay curve of an ideally energetically exponentially decaying impulse response is a function proportional to $e^{-t/\tau}$. In the following it is shown that this assumption is not valid for a superposition of decaying modes. The anti-derivative of the mono-frequent term can be expressed as follows by using Eq. A.4

$$
\begin{aligned}
L_{ii}(t) &= \frac{a_i^2 \tau_i}{4} \frac{1 + \tau_i^2 \omega_i^2 - \tau_i \omega_i \sin(2\omega_i t + 2\phi_i) + \cos(2\omega_i t + 2\phi_i)}{\tau_i^2 \omega_i^2 + 1} \cdot e^{-\frac{2t}{\tau_i}} \\
&= \frac{a_i^2 \tau_i}{4} \left(1 + \underbrace{\frac{\cos(2\omega_i t + 2\phi_i) - \tau_i \omega_i \sin(2\omega_i t + 2\phi_i)}{\tau_i^2 \omega_i^2 + 1}}_{\text{oscillating}} \right) \cdot e^{-\frac{2t}{\tau_i}}.
\end{aligned}
\tag{A.9}
$$

As can be seen from this equation the decay of the mono-frequent terms are mainly exponential as the oscillating component is much smaller than 1 and can therefore be neglected. For the mixed terms a different formulation is introduced as

$$
\begin{aligned}
L_{ij} &= \int_t^\infty 2a_i e^{-t'/\tau_i} \cos(\omega_i t' + \phi_i) a_j e^{-t'/\tau_j} \cos(\omega_j t' + \phi_j) \varepsilon(t') \, dt' \\
&= 4 \int_t^\infty a_i a_j e^{-\frac{t'}{\tau}} \left(\cos(\omega_\Sigma t' + \phi_i + \phi_j) + \cos(\omega_\Delta t' + \phi_i - \phi_j) \right) \varepsilon(t') \, dt'
\end{aligned}
\tag{A.10}
$$

with

$$
\begin{aligned}
\omega_\Sigma &= \omega_i + \omega_j \\
\omega_\Delta &= \omega_i - \omega_j \\
\frac{1}{\tau} &= \frac{1}{\tau_i} + \frac{1}{\tau_j}.
\end{aligned}
$$

This finally yields after integration with the corresponding integral limits

$$
\begin{aligned}
L_{ij} &= a_i a_j \tau \left(\frac{\cos(\omega_\Sigma t + \phi_i + \phi_j) - \tau \omega_\Sigma \sin(\omega_\Sigma t + \phi_i + \phi_j)}{\tau^2 \omega_\Sigma^2 + 1} \right) \cdot e^{-\frac{t}{\tau}} \\
&\quad + a_i a_j \tau \left(\frac{\cos(\omega_\Delta t + \phi_i - \phi_j) - \tau \omega_\Delta \sin(\omega_\Delta t + \phi_i - \phi_j)}{\tau^2 \omega_\Delta^2 + 1} \right) \cdot e^{-\frac{t}{\tau}}.
\end{aligned}
\tag{A.11}
$$

where this formulation can be rewritten in a more compact manner as

$$L_{ij} = a_i a_j \tau e^{-\frac{t}{\tau}} \left(\frac{\cos(\omega_\Sigma t + \phi_i + \phi_j + \phi)}{\sqrt{\tau^2 \omega_\Sigma^2 + 1}} + \frac{\cos(\omega_\Delta t + \phi_i - \phi_j + \phi)}{\sqrt{\tau^2 \omega_\Delta^2 + 1}} \right) \qquad (A.12)$$

with

$$\phi = -\arctan(\tau \omega_\Delta) \qquad (A.13)$$

As the function $L(t)$ is only valid for $t \geq 0$ the Heaviside function $\varepsilon(t)$ is eliminated by resetting the lower limit in the integral respectively. The second part of the equation is the potentially problematic part, e.g., in case two modes have almost the same eigenfrequency and amplitudes in the same order of magnitude. The frequency of this oscillating component can become very small. This causes the problem in the determination of the slope of the Schroeder curve.

A.3.2. Low-frequency Oscillation of the Schroeder Curve

As a basic example only two modes i and j are considered to show the effects of terms separately. All parameters for both modes are set to equal values except for the eigenfrequencies that are set to 100 Hz and 101 Hz. Figure A.3 illustrates the components separately and also the superposition of these components. As a cross-check the corresponding impulse response $h(t) = h_i + h_j$ is simulated and the calculated Schroeder curve of this impulse response yields the same results as the analytic formulation and hence the curves coincide in the given plot.

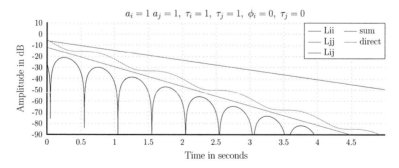

Figure A.3.: Schroeder curve for two room modes with the same amplitude, phase and decay but different eigenfrequencies of 100 Hz and 110 Hz.

As can be seen from Eq. A.11 there is a certain oscillation or ripple due to room modes with a small difference in the eigenfrequencies. The frequency f_{ripple} is determined by

$$f_{\text{ripple}} = |f_i - f_j| = \left|\frac{\omega_\Delta}{2\pi}\right|. \tag{A.14}$$

The amplitude of the ripple A_{ripple} can be determined by using the maximum and the minimum of $L(t)$. The maximum is given by the summation of the amplitudes of L_{ii}, L_{jj} and L_{ij}.

$$A_{\max} = \frac{a_i^2 \tau_i}{4} + \frac{a_j^2 \tau_j}{4} + \frac{a_i a_j \tau}{\tau^2 \omega_\Delta^2 + 1} \sqrt{1 + \tau^2 \omega_\Delta^2} \tag{A.15}$$

and the minimum as

$$A_{\min} = \frac{a_i^2 \tau_i}{4} + \frac{a_j^2 \tau_j}{4} - \frac{a_i a_j \tau}{\tau^2 \omega_\Delta^2 + 1} \sqrt{1 + \tau^2 \omega_\Delta^2}. \tag{A.16}$$

By linearizing the logarithm in a working point we can assume the logarithmic Schroeder curve to show sinusoidal ripple as well. The amplitude K_{ripple} of this ripple in $10 \log_{10}(L(t))$ can be written as half the difference between these amplitudes

$$K_{\text{ripple}} = 10 \log_{10}\left(\frac{A_{\max}}{A_{\min}}\right). \tag{A.17}$$

As mentioned before, this ripple becomes problematic for small frequency differences and if only a few modes interact. In room acoustic terms this means the low frequency range, e.g., below $1000\,\text{Hz}$ in domestic rooms. As the modal density increases over frequency, more and more modes interact and the ripples cancel each other out although the probability for very small differences in the eigenfrequencies rises as well.

In order to eliminate this oscillation and hence decreasing the uncertainty of the calculated reverberation time for different positions in the room, an approach using fitting of the impulse responses as discussed in Section 4.3.3 for the low frequency range and calculating the reverberation time by a weighted average of the modal decay rates could be investigated.

A.4. Changes in Temperature and Room Transfer Function

The temperature Θ has an important influence on the speed of sound c in air and is approximated in the following by

$$c = 331.3 \, \frac{\text{m}}{\text{s}} + 0.61 \frac{\text{m}}{\text{°C} \cdot \text{s}} \Theta \qquad (\text{A.18})$$

for typical temperatures observed in in-door and outdoor applications. This directly has an influence on the eigenfrequencies in the analytic model for the rectangular room in Eq. 4.13. The relative deviation of all eigenfrequencies is constant as they as their relation with the speed of sound is linear. It is interesting to investigate the deviations in the Frequency Response Function (FRF). The geometry L_1 is chosen exemplarily, but results can be also used to predict the deviations for the other geometries by scaling the frequency axis accordingly. The differences from the standard room temperature of $20°$C are chosen with increasing distances as $20.01°$C, $20.1°$C, $21°$C and $30°$C whereas the source and receiver position are chosen in opposite corners. The reverberation is kept constant for all temperatures and simulations are run up to $f_{\text{max}} = 20$ kHz.

These results can be used in different ways to calculate measures for this deviation. The suitable method of calculation depends on the purpose of this measure and it cannot be stated in a general way, which measure is preferable. The graphs are shown in Figure A.4 for RT $= 1$ s and $L_1 = (0.8 \, \text{m}, 0.5 \, \text{m}, 0.3 \, \text{m})$.

Commonly, the straightforward approach involving a division of the spectra (not smoothed, not averaged) is utilized as shown in Figure A.4(a). The peaks and dips deviating from the ideal straight line of 0 dB grow with increasing deviation in temperature. For the low frequency range around $200 - 400$ Hz the modal structure can be observed. With increasing frequency the modal density rises and the error increases at first whereas this relation is not linear.

The band levels of this error are shown in Figure A.4(b). With increasing deviation in temperature the errors increase as well, but a maximum occurs that moves towards lower frequencies.

For specific applications it is essential that the superposition approach is applicable. Hence, the error can be investigated by looking at the subtraction as shown in Figure A.4(c). As can be seen, the error is high at the peaks in the spectrum for the range of low modal density. The spectra are normalized to their maximum absolute value. Hence, the subtraction shows the attenuation of the error compared to the original result. This error increases over frequency. A simplified measure of this error is obtained by applying band averaging as shown in Figure A.4(d). As can be seen, the error increases gradually over frequency if the mode number is sufficiently high. An increase in temperature drift increases the errors especially for low frequencies. Similar results with the methods shown in Figure A.4(a) and Figure A.4(e) are published in (VORLÄNDER, 2013).

Finally, the difference in subtracting band levels of the original and the result with the deviating temperature is shown in Figure A.4(e). This clearly states, that the overall energetic spectrum can be obtained with even a large mismatch in temperature of 10 K.

The simplified measures are used to investigate the influence of the reverberation time on the error in Figure A.5. With increasing reverberation time the deviations measured by spectral division increase. For the subtraction the errors seem to decrease with increasing reverberation time. This is caused by the normalization, as the peaks in the spectrum for higher reverberation times increase. So this obtained error lies only further below the maximum values. It can be concluded that the observed error is dependent on frequency as it is dependent on the modal density and the mean reverberation time or modal decays. An increasing temperature variation directly leads to larger errors.

It is furthermore important to mention the missing dip around 250 Hz for the lowest reverberation time in Figure A.5(b) as the modal overlap is sufficiently high. Hence, a relation to the Schroeder frequency as commonly used in room acoustics could be found. Results shown here do not incorporate the correlation coefficient as shown for the exemplary room in Section 4.5.3.

A.5. Influence of the Size of the Scan Grid

Since the room acoustic measurements presented used only a limited scan area, the influence of this finite size is shortly investigated. Again, the source position is chosen to be in a corner. Furthermore, the scanned volume is chosen to be cubic with equal lengths starting in the opposite corner position. The size was gradually increased with the side lengths of 1 m, 2 m, 3 m and 4 m. The room geometry L_3 is chosen for this example.

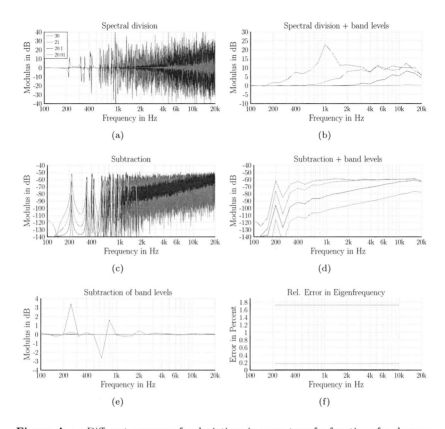

Figure A.4.: Different measures for deviations in room transfer functions for changes in temperature referenced to 20°C for RT = 1 s, geometry $L_1 =$ (0.8 m, 0.5 m, 0.3 m), and source and receiver in opposite corners.

Figure A.6 shows the observed deviation of the averaged modal coefficients over the distance. The limited scan area does not only influence the maximum distance that can be evaluated, but has also an effect on the result for the maximum distances. In case this maximum distance is still close to the distance that should be evaluated the result is overlayed by this effect. Ideally, all curves should coincide up to the point they are plotted to.

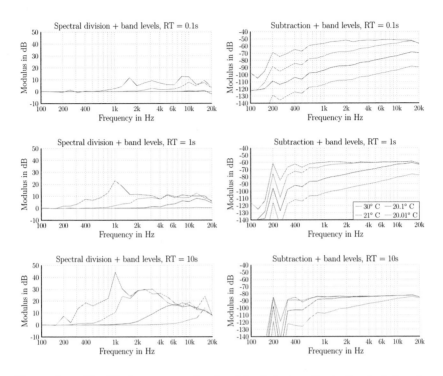

Figure A.5.: Simplified measures using band levels to investigate the deviations in the spectrum of the room transfer function using different reverberation times of 0.1 s (top), 1 s (center) and 10 s (low).

A.6. Influence of the Number of Modes

The analytic model of the rectangular room is used to investigate the influence of the number or modes on the shape of the deviation curve of the modal coefficients over the distance. In the model, the number of modes in a frequency can be manually decreased by discarding modes. 2000 random positions are simulated for different frequency bands. Results are depicted in Figure A.7. A strong dependence of the number of modes on the results cannot be observed. However, the results vary for the 63 Hz band, where an increasing number of modes shifts the results slightly away. The influence for the other frequency bands can mainly be observed at very large distances, where an increasing number of modes yields a more asymptotic behavior. This can be again explained by the averaging effect. The deviation over the normalized distance kd once again shows that the results

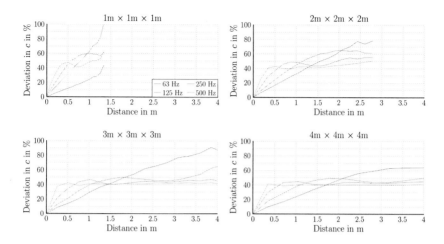

Figure A.6.: Influence on the size of the scanned volume on the observed deviation of the modal coefficient over distance.

in the different frequency bands are transferable. However, the finite spatial resolution caused by the limited number of observation points is observable for the two higher frequency bands. In contrast to the simulation used in Section 4.4.3 the number of simulation points is distributed over the entire volume and not a limited volume of a few cubic meters only. Hence, the minimum distance between observation points that could be evaluated is higher and hence the resolution is lower in the plots shown here. However, a settling of the deviation the coefficients over normalized distance can be stated to be around 3-4 independent on the number of modes per band.

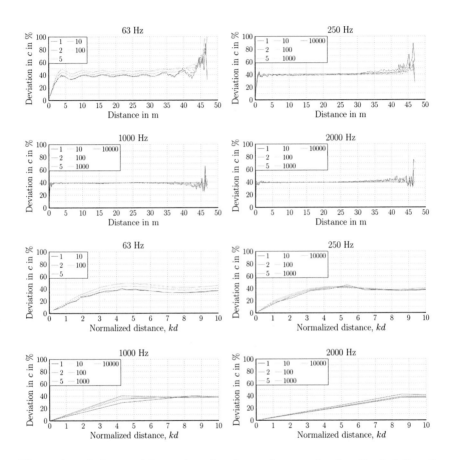

Figure A.7.: Influence of the number of modes per frequency band on the deviation of the modal coefficient over the distance. Absolute distance (top rows) and the normalized distance (bottom rows).

A.7. Implementation of OTPA Simulator

In order to investigate the influencing factors on the OTPA method, a simulation tool has been implemented in MATLAB using the ITA-Toolbox. This simulator is also under a BSD license. Additional helper routines have been added to the ITA-Toolbox in this context to realize matrix operations, inversions including regularization methods and the determination of physical units based on the input units. Furthermore, routines for the analytic plate model and to calculate the impedance coupling were implemented. The simulator itself is a documented MATLAB script (`ita_tutorial_OTPA_simulator`) and can be used as a tutorial for OTPA.

No measurement data is required and all parameters specifying the material constants can be defined in the simulator. It incorporates an analytic model for a simply supported plate (pinned-pinned) for the source and another simply supported plate with different geometries for the receiver as illustrated in Figure A.8. The analytic model is used to calculate the mobility matrices and the transfer functions by modal superposition up to maximum frequency specified where the boundary condition can also be specified for the source and receiver individually to clamped-clamped or free.

The source is characterized further by the position of its connection points, that are later connected to the receiver. Additionally, internal source positions are defined specifying the number of internal sources and the relation of their amplitudes at the connection points. The internal sources can be driven with arbitrary signals. The transfer path between these internal sources and the connection points is also obtained by the analytic model. Figure A.8(a) shows the source setup for a specific setting.

The receiver is modeled in a similar manner where the connection points should be chosen to match the geometry of the connections of the source but this is optional. With an arbitrary offset the source is placed on the receiver as depicted in Figure A.8(b). No radiation into air is calculated. The receiving point is chosen arbitrarily on the receiver plate again using the analytic model.

Between the source and the receiver vibration isolators can be used in the simulator. They are model using the two-port approach. Various scenarios using cross coupling between the isolator points or the degrees of freedom at one single isolator can be modeled. The basic model implemented uses a combination of a lumped parameter model motivation by a realistic scenario as published in (DIETRICH, HÖLLER, and LIEVENS, 2010).

Based on the input data further simplifications as described in Section 5.6 can be applied to investigate cases with, e.g., no cross coupling between the degrees of freedom that are theoretically not achievable with the plate model only. The block-size in samples for the OTPA Multiple Input Multiple Output (MIMO) method and the length of the entire source signal can be additionally controlled in order to investigate the influences on the obtained transfer functions. As the transfer functions are known by the model and are hence noiseless, the result of the OTPA can be further benchmarked against these reference results.

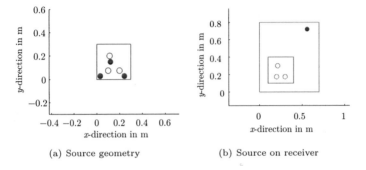

(a) Source geometry   (b) Source on receiver

Figure A.8.: Basic geometry used in the OTPA simulator based on an analytic model for simply supported plates of arbitrary size: a) source plate geometry, with internal source positions b) receiver plate geometry with receiving point (pink) and position of the source with the same connection points.

Bibliography

AES17 (1998). *AES standard method for digital audio engineering – Measurement of digital audio equipment – AES17-1998 (r2004).* Audio Engineering Society.

J. ALLEN and D. BERKLEY (1979). "Image method for efficiently simulating small-room acoustics." In: *Journal of the Acoustical Society of America (JASA)* 65.4, pp. 943–950.

M. ARETZ (2012). "Combined Wave And Ray Based Room Acoustic Simulations In Small Rooms—Challenges and limitations on the way to realistic simulation results." PhD thesis. Institute of Technical Acoustics, RWTH Aachen University.

M. ARETZ, P. DIETRICH, and M. VORLÄNDER ((2013)). "Application of the mirror source method for low frequency sound prediction in concave rooms." In: *Acta Acustica united with Acustica.* accepted.

G. BALLOU (1987). *Handbook for Sound Engineers.* Howard W. Sams.

S. R. BISTAFA and J. W. MORRISSEY (2003). "Numerical solutions of the acoustic eigenvalue equation in the rectangular room with arbitrary (uniform) wall impedances." In: *Journal of Sound and Vibration (JSV).*

J. BORISH and J. B. ANGELL (1983). "An Efficient Algorithm for Measuring the Impulse Response Using Pseudorandom Noiseo." In: *Journal of the Audio Engineering Society (JAES)* 31.

M. BOUCHARD, S. G. NORCROSS, and G. SOULODRE (2006). "Inverse Filtering Design Using a Minimal-Phase Target Function from Regularization." In: *121st AES Convention.* San Francisco, USA.

E. BRANDÃO, R. C. C. FLESCH, A. LENZI, and C. A. FLESCH (2011). "Estimation of pressure-particle velocity impedance measurement uncertainty using the Monte Carlo method." In: *Journal of the Acoustical Society of America (JASA).*

E. BRANDÃO, A. LENZI, and J. CORDIOLLI (2011). "Estimation and minimization of errors caused by sample size effect in the measurement of the normal absorption coefficient of a locally reactive surface." In: *Applied Acoustics.*

I. BRONSHTEIN, K. SEMENDYAYEV, G. MUSIOL, and H. MUEHLIG (2005). *Handbook of Mathematics.* Ed. by 5. Springer.

CEN/TS 1793-5 (2003). *Road traffic noise reducing devices – Test method for determining the acoustic performance – Part 5: Intrinsic characteristics - In situ values of sound reflection and airborne sound insulation.* ISO.

J.-P. CLAIRBOIS, F. de ROO, M. GARAI, M. CONTER, J. DEFRANCE, C. OLTEAN-DUMBRAVA, and I. FUSCO (2012). "QUIESST: third-term progress report." In: *Proceedings of Internoise*.

M. G. COX and B. R. L. SIEBERT (2006). "The use of a Monte Carlo method for evaluating uncertainty and expanded uncertainty." In: *Metrologia* 43.4, S178.

J. DAVY (1981). "The relative variance of the transmission function of a reverberation room." In: *Journal of Sound and Vibration (JSV)* 77.4, pp. 455–479.

D. DE KLERK and A. OSSIPOV (2010). "Operational transfer path analysis: Theory, guidelines and tire noise application." In: 24.7, pp. 1950–1962.

D. DESCHRIJVER, B. HAEGEMAN, and T. DHAENE (2007). "Orthonormal vector fitting: A robust macromodeling tool for rational approximation of frequency domain responses." In: *Advanced Packaging, IEEE Transactions on* 30.2, pp. 216–225.

J. DICKENS (1998). "Dynamic characterisation of vibration isolators." PhD thesis. Australia: University of New South Wales.

J. DICKENS (2002). "Review of Methods to Dynamically Represent Vibration Isolators." In: *The Shock and Vibration Digest* 34.6, p. 447.

J. DICKENS and C. NORWOOD (2001). "universal method to measure dynamic performance of vibration isolators under static load." In: *Journal of Sound and Vibration (JSV)* 244.4, pp. 685–696.

P. DIETRICH (Oct. 2007). "Measurement Uncertainties in Room Acoustics." Master's Thesis (Diplomarbeit). Aachen, Germany: Insitute of Technical Acoustics, RWTH Aachen University.

P. DIETRICH, M. ARETZ, M. MÜLLER-TRAPET, J. VAN GEMMEREN, and M. VORLÄNDER (Mar. 2012). "Uncertainty Factors in the Determination of Acoustic Reflection Factors with pu-probes." In: *38th German Annual Conference on Acoustics (DAGA)*. Darmstadt, Germany.

P. DIETRICH, M. GUSKI, J. KLEIN, M. MÜLLER-TRAPET, M. POLLOW, R. SCHARRER, and M. VORLÄNDER (Mar. 2013). "Measurements and Room Acoustic Analysis with the ITA-Toolbox for MATLAB." In: *40th Italian (AIA) Annual Conference on Acoustics and the 39th German Annual Conference on Acoustics (DAGA)*. Meran, Italy.

P. DIETRICH, M. GUSKI, M. POLLOW, B. MASIERO, M. MÜLLER-TRAPET, R. SCHARRER, and M. VORLÄNDER (Mar. 2012). "ITA-Toolbox – An Open Source MATLAB Toolbox for Acousticians." In: *38th German Annual Conference on Acoustics (DAGA)*. Darmstadt, Germany.

P. DIETRICH, M. GUSKI, and M. VORLÄNDER (Mar. 2013). "Influence of Loudspeaker Distortion on Room Acoustic Parameters." In: *40th Italian (AIA) Annual Conference on Acoustics and the 39th German Annual Conference on Acoustics (DAGA)*. Meran, Italy.

P. DIETRICH, C. HÖLLER, and M. LIEVENS (June 2010). "Simulation and Auralization of basic one-dimensional structure-borne sound sources using different modeling techniques." In: *39th International Congress and Exposition on Noise Control Engineering – Internoise*. Lisbon, Portugal.

P. DIETRICH, M. KUNKEMÖLLER, M. POLLOW, and M. VORLÄNDER (Mar. 2011). "Room Impulse Responses for Variable Source Radiation Patterns — Part 2: Application." In: *37th German Annual Conference on Acoustics (DAGA)*. Düsseldorf, Germany.

P. DIETRICH and M. LIEVENS (Mar. 2009). "How to Obtain High Quality Input Data for Auralization?" In: *International Conference on Acoustics and 35th German Annual Conference on Acoustics (NAG/DAGA)*. Rotterdam, Netherlands.

P. DIETRICH, M. LIEVENS, and S. PAUL (Nov. 2009). "Measurement Technique for High Quality Input Data for Auralization." In: *Proceedings of 20th International Congress of Mechanical Engineering (COBEM)*. Gramado, Rio Grande do Sul, Brazil.

P. DIETRICH, B. MASIERO, M. LIEVENS, and M. VORLÄNDER (Sept. 2010). "Open transfer path measurement round-robin using a simplified measurement object." In: *International Conference on Noise and Vibration Engineering (ISMA)*. Leuven, Belgium.

P. DIETRICH, B. MASIERO, and M. VORLÄNDER (Mar. 2013). "On the Optimization of the Multiple Exponential Sweep Method." In: *Journal of the Audio Engineering Society (JAES)* 61.3, pp. 113–124.

P. DIETRICH and I. WITEW (Mar. 2008). "Bewertung von Unsicherheitsbeiträgen bei raumakustischen Messungen nach GUM." In: *34th German Annual Conference on Acoustics (DAGA)*. Dresden, Germany.

C. DUNN and M. HAWKSFORD (1993). "Distortion immunity of MLS-derived impulse response measurements." In: *Journal of the Audio Engineering Society (JAES)* 41, pp. 314–314.

R. DURAISWAMI, D. ZOTKIN, and N. GUMEROV (2004). "Interpolation and range extrapolation of HRTFs [head related transfer functions]." In: *Acoustics, Speech, and Signal Processing, 2004. Proceedings.(ICASSP'04). IEEE International Conference on*. Vol. 4. IEEE, pp. iv–45.

A. ELLIOTT and A. MOORHOUSE (2010). "In-situ characterisation of structure borne noise from a building mounted wind turbine." In: *International Conference on Noise and Vibration Engineering (ISMA)*.

T. EVANS (2010). "Estimation of uncertainty in the structure- borne sound power transmission from a source to a receiver." PhD thesis. University of Salford.

F. J. FAHY (1995). "The vibro-acoustic reciprocity principle and applications to noise control." In: *Acustica* 81, p. 544.

F. FAHY and P. GARDONIO (2007). *Sound and Structural Vibration*. Academic Press.

F. FAHY and J. WALKER (2004). *Advanced Applications in Acoustics, Noise and Vibration*. spon press.

A. FARINA (2000). "Simultaneous measurement of impulse response and distortion with a swept-sine technique." In: *108th AES Convention*. Paris, France, pp. 18–22.

A. FARINA (2007). "Advancements in impulse response measurements by sine sweeps." In: *AES 122nd Convention*. Vol. 122. Vienna, Austria.

R. FELDTKELLER (1944). *Einführung in die Vierpoltheorie der elektrischen Nachrichtentechnik*. S. Hirzel.

P. Gajdatsy (2008a). "Critical Assessment of Operational Path Analysis: effect of coupling between inputs." In: *Euronoise*. Paris, France.

P. Gajdatsy (2008b). "Critical Assessment of Operational Path Analysis: Mathematical problems of transmissibility estimation." In: *Euronoise*. Paris, France.

P. A. Gajdatsy (2011). "Advanced Transfer Path Analysis Methods." PhD thesis. KU Leuven, Belgium.

P. Gajdatsy, K. Janssens, W. Desmet, and H. V. der Auweraer (2011). "Application of the Transmissibility Concept in Transfer Path Analysis." In: *International Conference on Noise and Vibration Engineering (ISMA)*. Leuven, Belgium.

P. Gajdatsy, K. Janssens, L. Gielen, P. Mas, and H. Van der Auweraer (2008). "Critical assessment of Operational Path Analysis: effect of neglected paths." In: *15th International Congress on Sound an Vibration*. Dajeon, Korea.

P. Gajdatsy, P. Sas, W. Desmet, K. Janssens, and H. Van der Auweraer (2011). "Effect of systematic FRF errors on matrix inversion based vibro-acoustic analysis methods." In: *Sensors, Instrumentation and Special Topics 2011: Proceedings of the 29th IMAC, a Conference on Structural Dynamics*. Vol. 6. Springer Verlag, p. 197.

M. Garai (2011). *Road traffic noise reducing devices – Noise reducing devices acting on airborne sound propagation – Test method for determining the acoustic performance – Intrinsic characteristics — In situ values of sound reflection under direct sound field conditions (Draft version 7) – D3.3 vr. 3.0*. Tech. rep. QUIESST.

A. Goertz (Nov. 2012). "Leistungsmessungen an Endstufen." In: *Production Partner*. Germany.

M. Guski, P. Dietrich, and M. Vorländer (Mar. 2012). "Positionsbedingte Unsicherheiten raumakustischer Parameter für geringe Modendichten anhand eines Reckteckraummodells." In: *38th German Annual Conference on Acoustics (DAGA)*. Darmstadt, Germany.

M. Guski and M. Vorländer (Mar. 2013). "Measurement Uncertainties of Reverberation Time caused by noise." In: *40th Italian (AIA) Annual Conference on Acoustics and the 39th German Annual Conference on Acoustics (DAGA)*. Meran, Italy.

B. Gustavsen and A. Semlyen (July 1999). "Rational approximation of frequency domain responses by vector fitting." In: *IEEE Transactions on Power Delivery* 14.3, pp. 1052 –1061.

Y. Haneda, S. Makino, Y. Kaneda, and N. Kitawaki (1999). "Common-acoustical-pole and zero modeling of head-related transfer functions." In: *IEEE Transactions on Speech and Audio Processing* 7.2, pp. 188–196.

Y. Haneda, S. Makino, and Y. Kaneda (Apr. 1994). "Common Acoustical Pole and Zero Modeling of Room Transfer Functions." In: *IEEE Transactions in Speech and Audio Processing* 2.2.

A. Hense (Nov. 2012). "Analise de incerteza em Salas Acusticas – Posicionamento da Fonte e do Receptor." MA thesis. Florianopolis, Brazil: LVA, UFSC.

K. Hirosawa, H. Nakagawa, M. Kon, and A. Yamamoto (2009). "Investigation of absorption coefficient measurement of acoustical materials by boundary element method using particle velocity and sound pressure sensor in free field." In: *Acoustical science and technology* 30.6, pp. 442–445.

C. HOELLER (May 2010). "Characterization of structure-borne sound sources in buildings." Master's Thesis (Diplomarbeit). Aachen, Germany: Institute of Technical Acoustics, RWTH Aachen University.

C. HUSZTY and S. SAKAMOTO (2010). "Time-domain sweeplets for acoustic measurements." In: *Applied Acoustics* 71.10, pp. 979–989.

P. HYNNÄ (2002). "vibrational power methods in control of sound and vibration." In: *Espoo: Technical Research Centre of Finland, VTT Industrial Systems.*

ISO 10846 (2005). *ISO 10846-1: Grundlagen und Übersicht (ISO/DIS 10846-1:2005).*

ISO 3382 (2009). *Acoustics – Measurement of Room Acoustic Parameters – Part 1.* ISO.

A. JANCZAK (2004). *Identification of nonlinear systems using neural networks and polynomial models: a block-oriented approach.* Vol. 310. Springer.

K. JANSSENS, P. GAJDATSY, L. GIELEN, P. MAS, L. BRITTE, W. DESMET, and H. VAN DER AUWERAER (2011). "OPAX: A new transfer path analysis method based on parametric load models." In: 25, pp. 1321–1338.

JCGM 100 (1995). *Guide to the Expression of Uncertainty in Measurement.* BIPM, IEC, IFCC, ILAC, ISO, IUPAC, IUPAP, OIML.

D. JIMENEZ, L. WANG, and Y. WANG (1991). "Linear quantization errors and the white noise hypothesis." In: *reconstruction* 10.

J. KEMP and H. PRIMACK (Dec. 2011). "Impulse Response and Measurement of Nonlinear and Systems: Properties of Existing Techniques and Wide Noise Sequences." In: *Journal of the Audio Engineering Society (JAES)* 12.59, pp. 953–963.

C. I. KLEIN (Sept. 2011). "Simulation, Verifizierung und Auralisierung der Körperschallsignale von Wälzlagern basierend auf Rauhigkeitsprofilen." Master's Thesis (Diplomarbeit). Aachen, Germany: Insitute for Electrical Drives (IEM), Institute of Technical Acoustics (ITA), and Werkzeugmaschinenlabor (WZL), RWTH Aachen University.

J. KLEIN (Mar. 2012). "Optimization of a Method for the Synthesis of Transfer Functions of Variable Sound Source Directivities for Acoustical Measurements." Master's Thesis (Diplomarbeit). Aachen, Germany: Institute of Technical Acoustics, RWTH Aachen University.

J. KLEIN, M. POLLOW, P. DIETRICH, and M. VORLÄNDER (Mar. 2013). "Room Impulse Response Measurements with Arbitrary Source Directivity." In: *40th Italian (AIA) Annual Conference on Acoustics and the 39th German Annual Conference on Acoustics (DAGA).* Meran, Italy.

W. KLIPPEL (1992). "Nonlinear large-signal behavior of electrodynamic loudspeakers at low frequencies." In: *Journal of the Audio Engineering Society (JAES)* 40.6, pp. 483–496.

T. KNÜTTEL, I. B. WITEW, and M. VORLÄNDER ((2013)). "Influence of "omnidirectional" loudspeaker directivity on measured room impulse responses." In: *Journal of the Acoustical Society of America (JASA).* submitted 03.10.2012, revised 21.04.2013.

M. KOB and M. VORLANDER (2000). "Band filters and short reverberation times." In: *Acta Acustica united with Acustica* 86.2, pp. 350–357.

M. KUNKEMÖLLER (Jan. 2011). "Entwicklung eines Analyse- und Syntheseverfahrens von mehrkanalig gemessenen Raumimpulsantworten für variable Quellrichtcharakteristiken." Master's Thesis (Diplomarbeit). Aachen, Germany: Institute of Technical Acoustics, RWTH Aachen University.

M. KUNKEMÖLLER, P. DIETRICH, and M. POLLOW (2011). "Synthesis of Room Impulse Responses for Variable Source Characteristics." In: *Acta Polytechnica – Journal of Advanced Engineering* 5.5, pp. 69–74.

M. KUNKEMÖLLER, M. POLLOW, P. DIETRICH, and M. VORLÄNDER (Mar. 2011). "Room Impulse Responses for Variable Source Radiation Patterns — Part 1: Synthesis." In: *37th German Annual Conference on Acoustics (DAGA)*. Düsseldorf, Germany.

H. KUTTRUFF (2000). *Room Acoustics*. Spon Press.

H. KUTTRUFF (2007). *Acoustics – An Introduction*. Taylor and Francis.

M. LIEVENS (2010). "Investigation Into the Importance of the Degrees of Freedom for the Characterisation of Structure-Borne Sound Sources." In: *Acta Acustica united with Acustica* 96.5, pp. 899–904.

M. LIEVENS (2013). "Structure-borne sound sources in buildings." PhD thesis. Institute of Technical Acoustics, RWTH Aachen University.

M. LIEVENS, C. HOELLER, P. DIETRICH, and M. VORLÄNDER ((2013)). "Predicting the interaction between structure-borne sound sources and receiver structures from independently measured quantities: Case study of a washing machine on a wooden joist floor." In: *Acta Acustica united with Acustica*. accepted.

S. LIPSHITZ, R. WANNAMAKER, and J. VANDERKOOY (Nov. 2004). "Dithered Noise Shapers and Recursive Digital Filters." In: *Journal of the Audio Engineering Society (JAES)* 52.11, pp. 1124–1141.

LMS (1995). *Transfer Path Analysis: The Qualification and Quantification of Vibro-Acoustic TransferPaths, LMS International, Application Note*. LMS.

LMS (2007). *Next generation Transfer Path Analysis to pinpoint noise and vibrationproblemsources–LMS Engineering Services pioneers new TPA techniquestoacceleratetroubleshootingthroughout the development process*. LMS.

P. MAJDAK, P. BALAZS, and B. LABACK (July 2007). "Multiple Exponential Sweep Method for Fast Measurement of Head-Related Transfer Functions." In: *Journal of the Audio Engineering Society (JAES)* 55.7-8, pp. 623–637.

R. S. MARTIN, I. WITEW, M. ARANA, and M. VORLANDER (2007). "Influence of the source orientation on the measurement of acoustic parameters." In: *Acta acustica united with acustica* 93.3, pp. 387–397.

B. MASIERO, R. F. BITENCOURT, P. DIETRICH, L. F. O. CHAMON, M. VORLÄNDER, and S. R. BISTAFA (May 2012). "Limiar Diferencial de Percepcao: um Estudo Sobre Respostas Impulsivas com Deslocamento do Receptor." In: *XXIV Encontro da Sobrac*. Belém, Pará, Brazil.

F. P. MECHEL (2008). *Formulas of Acoustics*. Springer.

R. MOERIG, F. J. BARR, D. L. NYLAND, and G. SITTON (Feb. 2004). "Method of using cascaded sweeps for source coding and harmonic cancellation." US Patent 6,687,619.

C. MOLLOY (1957). "Use of Four-Pole Parameters in Vibration Calculations." In: *Journal of the Acoustical Society of America (JASA)* 29, p. 842.

J. M. MONDOT and A. T. MOORHOUSE (2000). "Charcterisation of structure-borne sound sources: active componentsandassembledsystems." In: *Proceedings of Internoise*.

J. M. MONDOT and B. A. T. PETERSSON (1987). "Characterization of structure-borne sound sources-The source descriptor and the coupling function." In: *Journal of Sound and Vibration (JSV)* 114, pp. 507–518.

A. T. MOORHOUSE (2001). "On the characteristic power of structure-borne sound sources." In: *Journal of Sound and Vibration (JSV)* 248.3, pp. 441–459.

A. MOORHOUSE (printed 2007). "Simplified calculation of structure-borne sound from an active machinecomponentonasupporting substructure." In: *Journal of Sound and Vibration (JSV)* 302.1-2, pp. 67–87.

A. T. MOORHOUSE and B. M. GIBBS (1995). "Structure-borne sound power emission from resiliently mounted fans:casestudiesanddiagnosis." In: *Journal of Sound and Vibration (JSV)* 186.5, pp. 781–803.

J. MOURJOPOULOS and M. A. PARASKEVAS (1991). "Pole and zero modeling of room transfer functions." In: *Journal of Sound and Vibration* 146.2, pp. 281–302.

S. MÜLLER (1999). "Digitale Signalverarbeitung für Lautsprecher." PhD thesis. Aachen, Germany: Institute of Technical Acoustics, RWTH Aachen University.

S. MÜLLER and P. MASSARANI (2001). "Transfer-Function Measurement with Sweeps." In: *Journal of the Audio Engineering Society (JAES)* 49. printed, pp. 443–471.

S. MÜLLER and P. MASSARANI (2011). "Distortion immunity in impulse response measurements with sweeps." In: *ICSV 18*.

M. MÜLLER-TRAPET, P. DIETRICH, M. ARETZ, J. van GEMMEREN, and M. VORLÄNDER (Aug. 2013). "On the in-situ impedance measurement with pu-probes — simulation of the measurement setup." In: *Journal of the Acoustical Society of America (JASA)* 134.2, pp. 1082–1089.

Y. NAKA, A. A. OBERAI, and B. G. SHINN-CUNNINGHAM (2005). "Acoustic eigenvalues of rectangular rooms with arbitrary wall impedances using the interval NewtonŌgeneralized bisection method." In: *Journal of the Acoustical Society of America (JASA)* 118, pp. 3662–3671.

E. D. NELSON and M. L. FREDMAN (1970). "Hadamard Spectroscopy." In: *Journal of the Optical Society of America* 60.

K. NOUMURA and J. YOSHIDA (Oct. 2006a). "Method of transfer path analysis for vehicle interior sound with no excitation experiments." In: *Proc. FISITA 2006.* F2006D183 – 20068034. Yokohama (JP).

K. NOUMURA and J. YOSHIDA (2006b). "Method of Transfer Path Analysis for Interior Vehicle Sound Using Principal Component Regression Methods." In: *Honda R&D Tech Rev* 18.1, pp. 136–141.

A. NOVAK (2009). "Identification of Nonlinear Systems in Acoustics." PhD thesis. Universite du Maine, Le Mans, France.

A. NOVAK, L. SIMON, F. KADLEC, and P. LOTTON (Aug. 2010). "Nonlinear System Identification Using Exponential Swept-Sine Signal." In: *IEEE TRANSACTIONS ON INSTRUMENTATION AND MEASUREMENT* 59.8.

A. NOVAK, L. SIMON, and P. LOTTON (2010). "Analysis, Synthesis, and Classification of Nonlinear Systems Using Synchronized Swept-Sine Method for Audio Effects." In: *EURASIP Journal on Advances in Signal Processing.*

A. V. OPPENHEIM (1996). *Signals and Systems.* 2nd ed. Prentice-Hall.

T. OTSURU, R. TOMIKU, N. DIN, N. OKAMOTO, and M. MURAKAMI (2009). "Ensemble averaged surface normal impedance of material using an in-situ technique: Preliminary study using boundary element method." In: *Journal of the Acoustical Society of America (JASA)* 125, p. 3784.

B. A. T. PETERSSON and B. M. GIBBS (2000). "Towards a structure-borne sound source characterization." In: *Applied Acoustics* 61.3, pp. 325–343.

A. PIERCE (1989). *Acoustics: an introduction to its physical principles and applications.* Acoustical Society of America.

A. PIERSOL (1978). "Use Of Coherence And Phase Data Between Two Receivers In Evaluation Of Noise Environments." In: *Journal of Sound and Vibration* 56, pp. 215–228.

M. POLLOW (2007). "Variable Richtcharakteristik mit Dodekaeder-Lautsprechern – Variable directivity of dodecahedron loudspeakers." MA thesis. Aachen, Germany: Institute of Technical Acoustics, RWTH Aachen University.

M. POLLOW and G. K. BEHLER (Nov. 2009). "Variable Directivity for Platonic Sound Sources Based on Spherical Harmonics Optimization." In: *Acta Acustica united with Acustica* 95.6, pp. 1082–1092.

M. POLLOW, P. DIETRICH, B. KRECHEL, and M. VORLÄNDER (Mar. 2011). "Unidirektionale mehrkanalige Audioübertragung über Ethernet." In: *37th German Annual Conference on Acoustics (DAGA).* Düsseldorf, Germany.

M. POLLOW, P. DIETRICH, M. KUNKEMÖLLER, and M. VORLÄNDER (Oct. 2011). "Synthesis of room impulse responses for arbitrary source directivities using spherical harmonic decomposition." In: *IEEE Workshop on Applications of Signal Processing to Audio and Acoustics.* New Paltz, NY, USA.

M. POLLOW, P. DIETRICH, and M. VORLÄNDER (Mar. 2013). "Room Impulse Responses of Rectangular Rooms for Sources and Receivers of Arbitrary Directivity." In: *40th Italian (AIA) Annual Conference on Acoustics and the 39th German Annual Conference on Acoustics (DAGA).* Meran, Italy.

M. POLLOW, J. KLEIN, P. DIETRICH, G. K. BEHLER, and M. VORLÄNDER (Sept. 2012). "Optimized Spherical Sound Source for Room Reflections Analysis." In: *Proceedings of the International Workshop on Acoustic Signal Enhancement (IWAENG).* Aachen, Germany.

D. D. RIFE and J. VANDERKOOY (1989). "Transfer-Function Measurement with Maximum-Length Sequences." In: *Journal of the Audio Engineering Society (JAES)* 37.6, p. 419.

R. SOTTEK (2006). "Description of broadband structure-borne noise transmission from the power train using four-pole parameters." In: *Proceedings of Euronoise.*

Y. Suzuki, F. Asano, H.-Y. Kim, and T. Sone (1995). "An optimum computer-generated pulse signal suitable for the measurement of very long impulse responses." In: *The Journal of the Acoustical Society of America* 97, p. 1119.

A. N. Thite and D. J. Thompson (2003). "The quantification of structure-borne transmission paths by inverse methods.Part1:Improved singular value rejection methods." In: *Journal of Sound and Vibration (JSV)* 264.2, pp. 411–431.

M. Tohyama and T. Koike (1998). *Fundamentals of Acoustic Signal Processing.* Ac.

M. Tohyama and R. H. Lyon (1989). "Zeros of a transfer function in a multi-degree-of-freedom vibrating system." In: *Journal of the Acoustical Society of America (JASA)* 86. pole zero.

M. Tohyama, R. Lyon, and T. Koike (1994). "Phase variabilities and zeros in a reverberant transfer function." In: *Journal of the Acoustical Society of America (JASA)* 95, p. 286.

A. Torras-Rosell and F. Jacobsen (2010). "Measuring long impulse responses with pseudorandom sequences and sweep signals." In: *39th International Congress and Exposition on Noise Control Engineering – Internoise.*

H. Van der Auweraer, P. Mas, S. Dom, A. Vecchio, K. Janssens, and P. Van de Ponseele (2007). "Transfer Path Analysis in the Critical Path of Vehicle Refinement:TheRoleofFast, Hybrid and Operational Path Analysis." In: *SAE International* 2007-01-2352.

J. Vanderkooy (1994). "Aspects of MLS measuring systems." In: *Journal of the Audio Engineering Society (JAES)* 42.4, pp. 219–231.

J. Vanderkooy and S. P. Lipshitz (1984). "Resolution below the Least Significant Bit in Digital Systems with Dither." In: *Journal of the Audio Engineering Society (JAES)* 32.3.

R. Vermeulen, R. Lemmen, and A. B. J. Verheij (2001). "Active cancellation of unwanted excitation when measuring dynamicstiffnessofresilientelements." In: *Internoise.*

M. Vorländer (2007). *Auralization – Fundamentals of Acoustics, Modelling, Simulation, Algorithms and Acoustic Virtual Reality.* Springer Berlin.

M. Vorländer (2013). "Computer simulations in room acoustics: Concepts and uncertainties." In: *Journal of the Acoustical Society of America (JASA)* 133.3, p. 1203.

X. Wang ((2013)). "Model Based Signal Enhancement for Impulse Response Measurement." PhD thesis. ITA, RWTH Aachen University.

R. Wehr, M. Haider, M. Conter, S. Gasparoni, and S. Breuss (2013). "Measuring the sound absorption properties of noise barriers with inverse filtered maximum length sequences." In: *Applied Acoustics* 74.5, pp. 631–639.

S. Weinzierl, A. Giese, and A. Lindau (May 2009). "Generalized Multiple Sweep Measurement." In: *Audio Engineering Society Convention 126.*

E. G. Williams (1999). *Fourier Acoustics – Sound Radiation and Nearfield Acoustical Holography.* Academic Press.

I. Witew, P. Dietrich, S. Pelzer, and M. Vorländer (Mar. 2013). "Comparison of different modelling strategies to predict the spatial fluctuation of room acoustic single number quantities." In: *40th Italian (AIA) Annual Conference on Acoustics and the 39th German Annual Conference on Acoustics (DAGA)*. Meran, Italy.

I. WITEW, P. DIETRICH, D. de VRIES, and M. VORLÄNDER (Aug. 2010). "Uncertainty of room acoustic measurements — How many measurement positions are necessary to describe the conditions in auditoria?" In: *Proceedings of the International Symposium on Room Acoustics (ISRA)*. Melbourne, Australia.

I. WITEW, T. KNÜTTEL, and M. VORLÄNDER (2013). "Theoretic considerations on how the directivity of a sound source influences the measured impulse response." In: *Proceedings of Meetings on Acoustics*. Vol. 19, p. 015017.

I. WITEW and M. VORLÄNDER (2011). "Uncertainties of room acoustical measurements – influence of the exact source and receiver position." In: *Proceedings of the Institute of Acoustics*.

F. ZOTTER (2009). "Analysis and Synthesis of Sound-Radiation with Spherical Arrays." PhD thesis. Graz.

Publications

Journal Publications (with peer-review)

M. ARETZ, P. DIETRICH, and M. VORLÄNDER ((2013)). "Application of the mirror source method for low frequency sound prediction in concave rooms." In: *Acta Acustica united with Acustica*. accepted.

P. DIETRICH (2007). "Modeling Measurement Uncertainty in Room Acoustics." In: *Acta Polytechnica – Journal of Advanced Engineering* 47.4-5.

P. DIETRICH, B. MASIERO, and M. VORLÄNDER (Mar. 2013). "On the Optimization of the Multiple Exponential Sweep Method." In: *Journal of the Audio Engineering Society (JAES)* 61.3, pp. 113–124.

D. FRANCK, P. DIETRICH, M. van der GIET, K. HAMEYER, and M. VORLÄNDER (2012). "Simulation, Abstrahlung und Auralisation von elektrischen Maschinen als Hauptantrieb für die Kraftfahrzeuganwendung." In: *Lärmbekämpfung – Zeitschrift für Akustik, Schallschutz und Lärmbekämpfung, Springer VDI Verlag* 2, pp. 81–83.

M. KUNKEMÖLLER, P. DIETRICH, and M. POLLOW (2011). "Synthesis of Room Impulse Responses for Variable Source Characteristics." In: *Acta Polytechnica – Journal of Advanced Engineering* 5.5, pp. 69–74.

M. LIEVENS, C. HOELLER, P. DIETRICH, and M. VORLÄNDER ((2013)). "Predicting the interaction between structure-borne sound sources and receiver structures from independently measured quantities: Case study of a washing machine on a wooden joist floor." In: *Acta Acustica united with Acustica*. accepted.

B. MASIERO, M. POLLOW, P. DIETRICH, B. KRECHEL, and J. FELS ((2013)). "Design of a fast measurement setup for the range extrapolation of individual head-related transfer functions." In: *Acta Acustica united with Acustica*. accepted.

M. MÜLLER-TRAPET, P. DIETRICH, M. ARETZ, J. van GEMMEREN, and M. VORLÄNDER (Aug. 2013). "On the in-situ impedance measurement with pu-probes — simulation of the measurement setup." In: *Journal of the Acoustical Society of America (JASA)* 134.2, pp. 1082–1089.

M. MÜLLER-TRAPET, P. DIETRICH, and M. VORLÄNDER (May 2011). "Influence of Various Uncertainty Factors on the Result of Beamforming Measurements." In: *Noise Control Engineering Journal* 59.3, pp. 302–310.

Conference Publications (with peer-review)

P. DIETRICH, B. MASIERO, M. LIEVENS, and M. VORLÄNDER (Sept. 2010). "Open transfer path measurement round-robin using a simplified measurement object." In: *International Conference on Noise and Vibration Engineering (ISMA)*. Leuven, Belgium.

S. N. GERGES, R. F. BITENCOURT, C. SATO, P. DIETRICH, J. L. B. COELHO, and M. NEVES (Mar. 2008). "Comparison of Interior Jet Aircraft Noise using Sound Quality Parameters." In: *SAE Brazil NVH Conference*. Society of Automotive Engineers (SAE) International. Florianopolis, Brazil.

M. van der GIET, J. BLUM, P. DIETRICH, S. PELZER, M. MÜLLER-TRAPET, M. POLLOW, M. VORLÄNDER, and K. HAMEYER (May 2011). "Auralization of electrical in variable operating conditions." In: *IEEE International Electric Machines and Drives Conference (IEMDC)*. Niagara Falls, ON, Canada.

M. LIEVENS and P. DIETRICH (Apr. 2009). "Prediction of the sound radiation from a plate excited by a structure-borne sound source." In: *Noise and Vibration: Emerging Methods – NOVEM*. Oxford, UK.

S. PAUL and P. DIETRICH (Oct. 2009). "Quantifying slow amplitude and frequency modulations with psychoacoustic models – problems and preliminary solutions." In: *SAE, Brazil*. 2009-36-0357. Sao Paulo, Brazil.

M. POLLOW, P. DIETRICH, M. KUNKEMÖLLER, and M. VORLÄNDER (Oct. 2011). "Synthesis of room impulse responses for arbitrary source directivities using spherical harmonic decomposition." In: *IEEE Workshop on Applications of Signal Processing to Audio and Acoustics*. New Paltz, NY, USA.

C. T. SATO, M. L. ESPIGA, S. PAUL, and P. DIETRICH (May 2008). "Método analítico para o cálculo da Diferença Interaural de Tempo (ITD) no plano horizontal." In: *12th AES-Brazil National Convention and 6th AES-Brazil Conference*. Audio Engineering Society. Sao Paulo, Brazil.

Conference Publications

M. ARETZ, P. DIETRICH, and G. K. BEHLER (June 2010). "Comparison of in-situ measuring methods for absorption and surface impedances." In: *39th International Congress and Exposition on Noise Control Engineering – Internoise*. Invited Paper. Lisbon, Portugal.

G. K. BEHLER, P. DIETRICH, and M. VORLÄNDER (Mar. 2012a). "Mehrkanalmessungen an Lärmschutzwänden im Rahmen des QUIESST-Projekts." In: *38th German Annual Conference on Acoustics (DAGA)*. Darmstadt, Germany.

G. K. BEHLER, P. DIETRICH, and M. VORLÄNDER (Aug. 2012b). "Multi Channel Measurements for the Qualification of Noise Barriers In Situ, Discussion of Uncertainty Factors." In: *41st International Congress and Exposition on Noise Control Engineering – INTERNOISE*. Invited Paper. New York.

G. K. Behler, M. Pollow, P. Dietrich, and M. Vorländer (Aug. 2012). "Room impulse measurement and auralization with respect to source and receiver directivity." In: *41st International Congress and Exposition on Noise Control Engineering – INTERNOISE*. New York City, NY, USA.

J. Blum, M. Pollow, M. Müller-Trapet, M. van der Giet, P. Dietrich, K. Hameyer, and M. Vorländer (Mar. 2011). "Simulationskette zur Prädiktion der abgestrahlten Schallleistung elektrischer Maschinen." In: *37th German Annual Conference on Acoustics (DAGA)*. Düsseldorf, Germany.

P. Dietrich, M. Aretz, M. Müller-Trapet, J. van Gemmeren, and M. Vorländer (Mar. 2012). "Uncertainty Factors in the Determination of Acoustic Reflection Factors with pu-probes." In: *38th German Annual Conference on Acoustics (DAGA)*. Darmstadt, Germany.

P. Dietrich, M. Guski, J. Klein, M. Müller-Trapet, M. Pollow, R. Scharrer, and M. Vorländer (Mar. 2013). "Measurements and Room Acoustic Analysis with the ITA-Toolbox for MATLAB." In: *40th Italian (AIA) Annual Conference on Acoustics and the 39th German Annual Conference on Acoustics (DAGA)*. Meran, Italy.

P. Dietrich, M. Guski, M. Pollow, B. Masiero, M. Müller-Trapet, R. Scharrer, and M. Vorländer (Mar. 2012). "ITA-Toolbox – An Open Source MATLAB Toolbox for Acousticians." In: *38th German Annual Conference on Acoustics (DAGA)*. Darmstadt, Germany.

P. Dietrich, M. Guski, and M. Vorländer (Mar. 2013). "Influence of Loudspeaker Distortion on Room Acoustic Parameters." In: *40th Italian (AIA) Annual Conference on Acoustics and the 39th German Annual Conference on Acoustics (DAGA)*. Meran, Italy.

P. Dietrich, C. Höller, and M. Lievens (June 2010). "Simulation and Auralization of basic one-dimensional structure-borne sound sources using different modeling techniques." In: *39th International Congress and Exposition on Noise Control Engineering – Internoise*. Lisbon, Portugal.

P. Dietrich, M. Kunkemöller, M. Pollow, and M. Vorländer (Mar. 2011). "Room Impulse Responses for Variable Source Radiation Patterns — Part 2: Application." In: *37th German Annual Conference on Acoustics (DAGA)*. Düsseldorf, Germany.

P. Dietrich and M. Lievens (Mar. 2009). "How to Obtain High Quality Input Data for Auralization?" In: *International Conference on Acoustics and 35th German Annual Conference on Acoustics (NAG/DAGA)*. Rotterdam, Netherlands.

P. Dietrich, M. Lievens, and S. Paul (Nov. 2009). "Measurement Technique for High Quality Input Data for Auralization." In: *Proceedings of 20th International Congress of Mechanical Engineering (COBEM)*. Gramado, Rio Grande do Sul, Brazil.

P. Dietrich, B. Masiero, M. Müller-Trapet, M. Pollow, and R. Scharrer (Mar. 2010). "MATLAB Toolbox for the Comprehension of Acoustic Measurement and Signal Processing." In: *36th German Annual Conference on Acoustics (DAGA)*. Berlin, Germany.

P. Dietrich, B. Masiero, M. Pollow, B. Krechel, and M. Vorländer (Mar. 2012). "Time Efficient Measurement Method for Individual HRTFs." In: *38th German Annual Conference on Acoustics (DAGA)*. Darmstadt, Germany.

P. DIETRICH, B. MASIERO, R. SCHARRER, M. MÜLLER-TRAPET, M. POLLOW, and M. VORLÄNDER (Mar. 2011). "Application of the MATLAB ITA-Toolbox: Laboratory Course on Cross-talk Cancellation." In: *37th German Annual Conference on Acoustics (DAGA)*. Düsseldorf, Germany.

P. DIETRICH, B. MASIERO, and M. VORLÄNDER (Mar. 2010). "Vereinfachtes Messobjekt zur Untersuchung von Unsicherheitsfaktoren bei der Transferpfadanalyse und -synthese." In: *36th German Annual Conference on Acoustics (DAGA)*. Berlin, Germany.

P. DIETRICH and I. WITEW (Mar. 2008). "Bewertung von Unsicherheitsbeiträgen bei raumakustischen Messungen nach GUM." In: *34th German Annual Conference on Acoustics (DAGA)*. Dresden, Germany.

P. DIETRICH, I. WITEW, and M. VORLÄNDER (Mar. 2007). "Analyzing measurement uncertainty of room acoustic parameters." In: *33rd German Annual Conference on Acoustics (DAGA)*. Stuttgart, Germany.

S. FINGERHUTH, P. DIETRICH, and R. KALDENBACH (Mar. 2010). "Mess-'Blackbox' zum Verständnis des Übertraungsverhaltens und der akustischen Messtechnik." In: *36th German Annual Conference on Acoustics (DAGA)*. Berlin, Germany.

S. FINGERHUTH, P. DIETRICH, M. POLLOW, M. VORLÄNDER, D. FRANCK, M. van der GIET, K. HAMEYER, M. BÖSING, K. A. KASPER, and R. W. D. DONCKER (Sept. 2009). "Towards the auralization of electrical machines in complex virtual scenarios." In: *40th National Congress on Acoustics –Tecniacústica*. Cádiz, Spain.

W. D. FONSECA, B. MASIERO, S. BISTAFA, P. DIETRICH, G. QUIQUETO, L. CHAMON, and M. VORLÄNDER (May 2010). "Medição de uma plataforma acústica conceitual desenvolvida por diferentes instituições de pesquisa." In: *XXIII Encontro da Sociedade Brasileira de Acústica (SOBRAC)*. SOBRAC. Salvador, Bahia, Brazil.

D. FRANCK, M. van der GIET, P. DIETRICH, K. HAMEYER, and M. VORLÄNDER (Nov. 2011). "Analysis, auralization and reduction of electromagnetic excited audible noise for electrical vehicles." In: *Aachen Acoustic Colloquium (AAC)*. Aachen, Germany.

M. van der GIET, M. MÜLLER-TRAPET, P. DIETRICH, M. POLLOW, J. BLUM, K. HAMEYER, and M. VORLÄNDER (June 2010). "Comparison of acoustic single-value parameters for the design process of electrical machines." In: *39th International Congress and Exposition on Noise Control Engineering – Internoise*. Lisbon, Portugal.

M. GUSKI, P. DIETRICH, and M. VORLÄNDER (Mar. 2012). "Positionsbedingte Unsicherheiten raumakustischer Parameter für geringe Modendichten anhand eines Reckteckraummodells." In: *38th German Annual Conference on Acoustics (DAGA)*. Darmstadt, Germany.

J. KLEIN, P. DIETRICH, M. POLLOW, and M. VORLÄNDER (Mar. 2012). "Optimized Measurement System for the Synthesis of Transfer Functions of Variable Sound Source Directivities for Acoustical Measurements." In: *38th German Annual Conference on Acoustics (DAGA)*. Darmstadt, Germany.

J. KLEIN, M. POLLOW, and P. DIETRICH (May 2012). "Optimized System for the Synthesis of Optimized System for the Synthesis of Room Impulse Responses of Arbitrary Sound Sources." In: *POSTER 2012: 16th International Student Conference on Electrical Engineering*. Prague, Czech Republic.

J. KLEIN, M. POLLOW, P. DIETRICH, and M. VORLÄNDER (Mar. 2013). "Room Impulse Response Measurements with Arbitrary Source Directivity." In: *40th Italian (AIA) Annual Conference on Acoustics and the 39th German Annual Conference on Acoustics (DAGA)*. Meran, Italy.

B. KRECHEL, P. DIETRICH, M. POLLOW, and B. MASIERO (May 2012). "Fast Measurements of Individual HRTFs Using Multiple Exponential Sweeps." In: *POSTER 2012: 16th International Student Conference on Electrical Engineering*. Prague, Czech Republic.

M. KUNKEMÖLLER, P. DIETRICH, and M. POLLOW (May 2011). "Synthesis of Room Impulse Responses for Variable Source Characteristics." In: *POSTER 2012: 16th International Student Conference on Electrical Engineering*. Prague, Czech Republic.

M. KUNKEMÖLLER, P. DIETRICH, M. POLLOW, and M. VORLÄNDER (Mar. 2012). "Synthesis of Room Impulse Response for Arbitrary Source Directivities using Spherical Harmonic Decomposition." In: *38th German Annual Conference on Acoustics (DAGA)*. Darmstadt, Germany.

M. KUNKEMÖLLER, M. POLLOW, P. DIETRICH, and M. VORLÄNDER (Mar. 2011). "Room Impulse Responses for Variable Source Radiation Patterns — Part 1: Synthesis." In: *37th German Annual Conference on Acoustics (DAGA)*. Düsseldorf, Germany.

B. MASIERO, R. F. BITENCOURT, P. DIETRICH, L. F. O. CHAMON, M. VORLÄNDER, and S. R. BISTAFA (May 2012). "Limiar Diferencial de Percepcao: um Estudo Sobre Respostas Impulsivas com Deslocamento do Receptor." In: *XXIV Encontro da Sobrac*. Belém, Pará, Brazil.

B. MASIERO, P. DIETRICH, M. POLLOW, J. FELS, and M. VORLÄNDER (Mar. 2012). "Design of a Fast Individual HRTF Measurement System." In: *38th German Annual Conference on Acoustics (DAGA)*. Darmstadt, Germany.

B. S. MASIERO, W. D. FONSECA, M. MÜLLER-TRAPET, and P. DIETRICH (Sept. 2010). "Auralization of pass-by beamforming measurements." In: *EAA Euroregio Congress on Sound and Vibration*. EAA. Ljubljana, Slovenia.

M. MÜLLER-TRAPET and P. DIETRICH (Mar. 2009). "Comparison of sound-source localization techniques for vibrating structures." In: *International Conference on Acoustics and 35th German Annual Conference on Acoustics (NAG/DAGA)*. Rotterdam, Netherlands.

M. MÜLLER-TRAPET and P. DIETRICH (Apr. 2010). "Influence of Various Uncertainty Factors on the Result of Beamforming Measurements." In: *25th conference of the Institute of Noise Control Engineering – NoiseCon 2010*. Baltimore, Maryland, USA.

M. MÜLLER-TRAPET, P. DIETRICH, M. van der GIET, J. BLUM, M. VORLÄNDER, and K. HAMEYER (Sept. 2010). "Simulated Transfer Functions for the Auralization of Electrical Machines." In: *EAA Euroregio Congress on Sound and Vibration*. EAA. Ljubljana, Slovenia.

S. PAUL and P. DIETRICH (Oct. 2011). "Measurements of electrical transfer characteristics of soundcards as classroom activity." In: *162nd Meeting of the Acoustical Society of America*. 2aEDa11. no paper. ASA. San Diego, California, USA.

S. PAUL, P. DIETRICH, J. J. de SOUZA, and M. VORLÄNDER (Mar. 2011). "Einsatz der ITA-Toolbox in einem Grundlagenkurs zur Signalverarbeitung." In: *37th German Annual Conference on Acoustics (DAGA)*. Düsseldorf, Germany.

A. Pedrero, M. Pollow, P. Dietrich, G. Behler, M. Vorländer, C. Díaz, and A. Díaz (Oct. 2012). "Mozarabic Chant anechoic recordings for auralization purposes." In: *43rd National Congress on Acoustics - Tecniacústica*. Evora, Portugal.

M. Pollow, P. Dietrich, B. Krechel, and M. Vorländer (Mar. 2011). "Unidirektionale mehrkanalige Audioübertragung über Ethernet." In: *37th German Annual Conference on Acoustics (DAGA)*. Düsseldorf, Germany.

M. Pollow, P. Dietrich, B. Masiero, J. Fels, and M. Vorländer (Mar. 2012). "Modal sound field representation of HRTFs." In: *38th German Annual Conference on Acoustics (DAGA)*. Darmstadt, Germany.

M. Pollow, P. Dietrich, and M. Vorländer (Mar. 2013). "Room Impulse Responses of Rectangular Rooms for Sources and Receivers of Arbitrary Directivity." In: *40th Italian (AIA) Annual Conference on Acoustics and the 39th German Annual Conference on Acoustics (DAGA)*. Meran, Italy.

M. Pollow, J. Klein, P. Dietrich, G. K. Behler, and M. Vorländer (Sept. 2012). "Optimized Spherical Sound Source for Room Reflections Analysis." In: *Proceedings of the International Workshop on Acoustic Signal Enhancement (IWAENG)*. Aachen, Germany.

M. Pollow, B. Masiero, P. Dietrich, J. Fels, and M. Vorländer (Oct. 2012). "Fast measurement system for spatially continuous individual HRTFs." In: *Audio Engineering Society 25th UK Conference in association with the 4th International Symposium on Ambisonics and Spherical Acoustics*. London, UK.

J. J. de Souza, S. Paul, E. Brandão, and P. Dietrich (Sept. 2013). "Comparison of simulations and measurements for a simplified acoustic enclosure." In: *42nd International Congress and Exposition on Noise Control Engineering*. accepted. Innsbruck, Austral.

M. Spiertz, V. Gnann, P. Dietrich, J.-R. Ohm, and M. Vorländer (Mar. 2011). "Single Sensor Source Separation for Acoustical Machine Diagnostics." In: *37th German Annual Conference on Acoustics (DAGA)*. Düsseldorf, Germany.

I. Witew and P. Dietrich (Sept. 2007). "Assessment of the uncertainty in room acoustical measurements." In: *19th International Congress on Acoustics*. Sociedad Espanola de Acustica (SEA). Madrid, Spain.

I. Witew, P. Dietrich, S. Pelzer, and M. Vorländer (Mar. 2013). "Comparison of different modelling strategies to predict the spatial fluctuation of room acoustic single number quantities." In: *40th Italian (AIA) Annual Conference on Acoustics and the 39th German Annual Conference on Acoustics (DAGA)*. Meran, Italy.

I. Witew, P. Dietrich, and M. Vorländer (Aug. 2010). "Error and uncertainty of IACC measurements introduced by dummy head orientation using Monte Carlo simulations." In: *Proceedings of 20th International Congress on Acoustics (ICA)*. Australian Acoustical Society. Sydney, Australia.

I. Witew, P. Dietrich, D. de Vries, and M. Vorländer (Aug. 2010). "Uncertainty of room acoustic measurements — How many measurement positions are necessary to describe the conditions in auditoria?" In: *Proceedings of the International Symposium on Room Acoustics (ISRA)*. Melbourne, Australia.

Supervised Thesis

Supervised Bachelor Theses

H. Behm (Apr. 2010). "Konstruktion und Erprobung eines Kalibriersystems für Elektretmikrofone." Bachelor's Thesis (Studienarbeit). Aachen, Germany: Institute of Technical Acoustics, RWTH Aachen University.

C. Haar (Sept. 2012). "Evaluation of Binaural Reproduction Techniques with In-Ear Systems." Bachelor's Thesis. Aachen, Germany: Institute of Technical Acoustics, RWTH Aachen University.

C. Haentjes (Nov. 2010). "Design und Entwicklung einer GUI für in-situ Impedanzmessverfahren." Bachelor's Thesis (Studienarbeit). Aachen, Germany: Institute of Technical Acoustics, RWTH Aachen University.

A. Kludszuweit (Apr. 2012). "Nichtlineare Modellierung und Messung von akustischen Systemen." Bachelor's Thesis (Studienarbeit). Aachen, Germany: Institute of Technical Acoustics, RWTH Aachen University.

B. Krechel (Oct. 2011). "Unidirektionale mehrkanalige Audioübertragung über Ethernet." Bachelor's Thesis. Aachen, Germany: Institute of Technical Acoustics, RWTH Aachen University.

J. Nahas (Jan. 2010). "Messung der Lautsprecherrichtcharakteristik für akustische Maßstabsmodelle." Bachelor's Thesis (Studienarbeit). Aachen, Germany: Institute of Technical Acoustics, RWTH Aachen University.

J. Tumbrägel (May 2013). "Implementation and Evaluation of 3D Monitoring for Musicians." Bachelor's Thesis. Institute of Technical Acoustics, RWTH Aachen University.

Supervised Master Theses

A. Bleus (Aug. 2010). "Nonlinearities in Loudspeakers – Measurements and Models." Master's Thesis. Aachen, Germany: Institute of Technical Acoustics, RWTH Aachen University.

J. BLUM (Sept. 2010). "Development and Implementation of a Procedure for the Fast Auralization of Rotating Electrical Machines under Variable Operation Conditions." Master's Thesis (Diplomarbeit). Aachen, Germany: Institute of Technical Acoustics, RWTH Aachen University.

D. CRAGG (Mar. 2011). "Psychoakustische Bewertung von Rechenmethoden zur schnellen Auralisation elektrischer Maschinen." Master's Thesis (Diplomarbeit). Aachen, Germany: Institute of Technical Acoustics, RWTH Aachen University.

J. van GEMMEREN (Sept. 2011). "Quantitative Bewertung von Unsicherheitsfaktoren bei der In-Situ Messung akustischer Reflexionsfaktoren mit einem Microflown pu-Sensor." Master's Thesis (Diplomarbeit). Aachen, Germany: Institute of Technical Acoustics, RWTH Aachen University.

A. HENSE (Nov. 2012). "Analise de incerteza em Salas Acusticas – Posicionamento da Fonte e do Receptor." MA thesis. Florianopolis, Brazil: LVA, UFSC.

C. HOELLER (May 2010). "Characterization of structure-borne sound sources in buildings." Master's Thesis (Diplomarbeit). Aachen, Germany: Institute of Technical Acoustics, RWTH Aachen University.

M. HORN (Dec. 2007). "Entwicklung eines Prototypen zur nicht-invasiven Messung der Schwingspulentemperatur von Lautsprechern." MA thesis. Hochschule für Technik und Wirtschaft Dresden and Institute of Technical Acoustics, RWTH Aachen University.

R. KENGNI (2013). "Automatische Optimierung der akustischen Messsystemaussteuerung mit Sweep-Messtechnik – Automatic dynamic range optimization of sweep measurements." MA thesis. Institute of Technical Acoustics, RWTH Aachen University.

C. I. KLEIN (Sept. 2011). "Simulation, Verifizierung und Auralisierung der Körperschallsignale von Wälzlagern basierend auf Rauhigkeitsprofilen." Master's Thesis (Diplomarbeit). Aachen, Germany: Insitute for Electrical Drives (IEM), Institute of Technical Acoustics (ITA), and Werkzeugmaschinenlabor (WZL), RWTH Aachen University.

J. KLEIN (Mar. 2012). "Optimization of a Method for the Synthesis of Transfer Functions of Variable Sound Source Directivities for Acoustical Measurements." Master's Thesis (Diplomarbeit). Aachen, Germany: Institute of Technical Acoustics, RWTH Aachen University.

B. KRECHEL (Mar. 2012). "Schnelle Messung von individuellen HRTFs mit kontinuierlichen MIMO-Verfahren." Master's Thesis. Aachen, Germany: Institute of Technical Acoustics, RWTH Aachen University.

M. KUNKEMÖLLER (Jan. 2011). "Entwicklung eines Analyse- und Syntheseverfahrens von mehrkanalig gemessenen Raumimpulsantworten für variable Quellrichtcharakteristiken." Master's Thesis (Diplomarbeit). Aachen, Germany: Institute of Technical Acoustics, RWTH Aachen University.

M. MÜLLER-TRAPET (Apr. 2009). "Comparison of Sound-Source Localization Methods for Vibrating Structures." Master's Thesis (Diplomarbeit). Aachen, Germany: Institute of Technical Acoustics, RWTH Aachen University.

M. PRAAST (Oct. 2009). "Optimierung und Vergleich unterschiedlicher Verfahren zur Messung der akustischen Impedanz." Master's Thesis (Diplomarbeit). Aachen, Germany: Institute of Technical Acoustics, RWTH Aachen University.

Curriculum Vitae

Personal Data

Pascal Dietrich

4. 12. 1981   born in Dorsten, Germany

Education

1992 – 2001   Gymnasium Petrinum Dorsten, Germany

1988 – 1992   St. Antonius Grundschule, Dorsten, Germany

Higher Education

10/2001 – 10/2006   Masters' Degree in Electrical Engineering (Dipl.-Ing.)
RWTH Aachen University
Focus: Communication- and Information Technology

Professional Experience

02/2007 – today   Member of the research staff at
Institute of Technical Acoustics (ITA), RWTH Aachen

02/2008 – 08/2008 Researcher at Laboratório de Vibrações e Acústica (LVA),
Universidade Federal de Santa Catarina (UFSC), Flori-
anópolis, Santa Catarina, Brazil

09/2005 – 02/2006 Intern at Siemens Corporate Research,
Princeton, New Jersey, USA

Aachen, Germany, November 6, 2013

Bisher erschienene Bände der Reihe

Aachener Beiträge zur Technischen Akustik

ISSN 1866-3052

Alle erschienenen Bücher können unter der angegebenen ISBN-Nummer direkt online (http://www.logos-verlag.de) oder per Fax (030 - 42 85 10 92) beim Logos Verlag Berlin bestellt werden.